Sustainable Development Goals Series

The **Sustainable Development Goals Series** is Springer Nature's inaugural cross-imprint book series that addresses and supports the United Nations' seventeen Sustainable Development Goals. The series fosters comprehensive research focused on these global targets and endeavours to address some of society's greatest grand challenges. The SDGs are inherently multidisciplinary, and they bring people working across different fields together and working towards a common goal. In this spirit, the Sustainable Development Goals series is the first at Springer Nature to publish books under both the Springer and Palgrave Macmillan imprints, bringing the strengths of our imprints together.

The Sustainable Development Goals Series is organized into eighteen subseries: one subseries based around each of the seventeen respective Sustainable Development Goals, and an eighteenth subseries, "Connecting the Goals," which serves as a home for volumes addressing multiple goals or studying the SDGs as a whole. Each subseries is guided by an expert Subseries Advisor with years or decades of experience studying and addressing core components of their respective Goal.

The SDG Series has a remit as broad as the SDGs themselves, and contributions are welcome from scientists, academics, policymakers, and researchers working in fields related to any of the seventeen goals. If you are interested in contributing a monograph or curated volume to the series, please contact the Publishers: Zachary Romano [Springer; zachary.romano@springer.com] and Rachael Ballard [Palgrave Macmillan; rachael.ballard@palgrave.com].

More information about this series at
https://link.springer.com/bookseries/15486

Parul Rishi

Managing Climate Change and Sustainability through Behavioural Transformation

palgrave
macmillan

Parul Rishi
Faculty of Human Resource Management
Indian Institute of Forest Management
Bhopal, Madhya Pradesh, India

ISSN 2523-3084　　　　　ISSN 2523-3092　(electronic)
Sustainable Development Goals Series
ISBN 978-981-16-8518-7　　　ISBN 978-981-16-8519-4　(eBook)
https://doi.org/10.1007/978-981-16-8519-4

The content of this publication has not been approved by the United Nations and does not reflect the views of the United Nations or its officials or Member States.

© The Editor(s) (if applicable) and The Author(s), under exclusive license to Springer Nature Singapore Pte Ltd. 2022
Color wheel and icons: From https://www.un.org/sustainabledevelopment/, Copyright © 2020 United Nations. Used with the permission of the United Nations.
This work is subject to copyright. All rights are solely and exclusively licensed by the Publisher, whether the whole or part of the material is concerned, specifically the rights of translation, reprinting, reuse of illustrations, recitation, broadcasting, reproduction on microfilms or in any other physical way, and transmission or information storage and retrieval, electronic adaptation, computer software, or by similar or dissimilar methodology now known or hereafter developed.
The use of general descriptive names, registered names, trademarks, service marks, etc. in this publication does not imply, even in the absence of a specific statement, that such names are exempt from the relevant protective laws and regulations and therefore free for general use.
The publisher, the authors and the editors are safe to assume that the advice and information in this book are believed to be true and accurate at the date of publication. Neither the publisher nor the authors or the editors give a warranty, expressed or implied, with respect to the material contained herein or for any errors or omissions that may have been made. The publisher remains neutral with regard to jurisdictional claims in published maps and institutional affiliations.

Cover illustration: S_Lew/Getty Images/iStockphoto

This Palgrave Macmillan imprint is published by the registered company Springer Nature Singapore Pte Ltd.
The registered company address is: 152 Beach Road, #21-01/04 Gateway East, Singapore 189721, Singapore

This book is dedicated to

*THE GOD Almighty for HIS divine grace
to grant wisdom and ignite passion for perennial learning*

and my loving parents

Smt. Vijay Sethia and Shri. Partap Rai Sethia

*for their untiring lifetime support
to realize my dreams and make me what I am today.*

Foreword

The book 'Managing Climate Change and Sustainability through Behavioural Transformation' has been written from holistic perspective, that takes into consideration the important role of transforming human behaviour for global sustainability and the better health of the planet, as part of our efforts to handle climates such as climate change. It is a very useful source of information for its readers on issues related to climate change risk appraisal, perceptions, coping and adaptation, contemplative practices, and pro-climate action. The book has the potential to encourage its readers to not only understand the importance of behaving responsibly, but also make attempts to meaningfully engage themselves with sustainable behavioural practices.

Human fingerprints are all over the planet. There is certainly not much that we can do, as humans, to undo the damage that has already been caused to the planet, but we certainly can make attempts at the country, local and individual levels, to make sure that the impacts of human activities do not go beyond the already existing 'anthropogenic contributions'. By way of this book, the author has succeeded in her efforts to highlight insightful behavioural suggestions and solutions which may go a long way in helping its readers make wise and informed behavioural choices. It is only when people understand the ways to connect their everyday energy usage, disposal of waste, actions-purchases and other activities with the larger concepts of climate change, we may make some progress towards reducing GHG emissions and the subsequent ecological impacts.

A special mention may be made towards the end of the book, where interesting topics like 'Frugality and innovations', 'contemplative practices', 'mindfulness' among several others, are tackled. This may not only give its readers a very satisfying reading experience but also inspiration and motivation to act and behave responsibly because we don't have a planet 'B', after all !

Walter Leal
Professor of Environment and
Technology
Manchester Metropolitan
University
Manchester, UK

Preface

In today's fast progressing world with technology at the forefront, we all have seen the 'Urban Trance', hastily moving towards urbanization and industrialization. At the same time, we have also experienced the hard reality of changing climate across the globe, due to varied causes, but primarily, physical and anthropogenic in nature. However, the way globalization and urbanization are differentially instrumental in causing climatic adversities to different regions of planet earth and posing diversified challenges to sustainability management, is an issue which requires extensive debate at the global front. Positioning climate change and sustainability in a system-based perspective makes us understand the fact that focus on specific elements may not be sufficient to understand the concept in a 'holistic' perspective. No single discipline can account for the dynamics of sustainability behaviour and climate change. In this perspective, a critical analysis of human behaviour in the context of climate change and sustainability management, is pertinent for wider policy implications.

The motivation for writing this book was derived from my almost three-decade old association as a student of behavioural science. My constant search to look for its uniquely positioned applications in environment, natural resource management, sustainability and climate change ignited my thoughts, which got structured in the form of this volume. The book will show how psychology can uniquely contribute to the understanding of climate change and how behavioural processes are crucial to be associated with the process of sustainability management. It further talks about integration of Corporate Social Responsibility (CSR) practices in the backdrop of Sustainable Development Goals (SDGs) to connect them to principles of sustainability as a practically useful, contextually relevant and research-based text.

Bhopal, India   Parul Rishi

Acknowledgements

Completion of this book was possible with the support and contributions of many people, to whom I want to acknowledge here.

First and the foremost, I want to humbly acknowledge my mentor, Prof Ramesh. K. Arora, chairman, Management Development Academy, for evaluating and considering this thematic proposal worthwhile and useful for the readers. His constant motivation and guidance at different stages of this book was the source behind its timely completion. Further, I also acknowledge the constant encouragement and support of my senior colleague Dr. Suprava Patnaik of Indian Institute of Forest Management, Bhopal, for helping me to withstand all odds and take this book towards a logical conclusion. Sincere acknowledgement is also due to Dr. Vidhya Sagar Athota, University of Notre Dame, Australia, for motivating me to take up this book and his valuable academic and functional contributions during initial stages of this book. I am also thankful to Mr. Prem Prakash Srivastava for his contributions regarding documentation of Dayalbagh Eco-City and SPHEEHA activities.

Academic and editorial support at different stages of this book, constantly extended by my scholars- Dr. Ruchi Mudaliar, Dr. Nidhi Sharma, Mr. Pavan Balakrishna, Ms. Soumya Gupta and Ms. Shalini Dagur also deserve due acknowledgement. Their untiring support was my biggest strength.

This book was written during globally challenging times. My constant source of strength and motivation was my husband, Dr. Rishi Nigam and daughters, Tarana and Siddhi, who embraced my dreams passionately and were always there in all odds and shines of my life while writing this book.

Finally, I am thankful to my Alma Matter and temple of my learning—Dayalbagh University, Agra, for igniting the constant quest for learning and Indian Institute of Forest Management, Bhopal, India, for providing me conducive academic environment, to complete this book.

PRAISE FOR *MANAGING CLIMATE CHANGE AND SUSTAINABILITY THROUGH BEHAVIOURAL TRANSFORMATION*

"'Managing Climate Change and Sustainability through Behavioural Transformation' is one of those rare books which demonstrates that behavioural aspects are as important as technology for creating a sustainable future. The book addresses the issue of climate change and sustainability from a very pertinent but often less-addressed viewpoint of inculcating behavioural changes as a means of orienting the global society towards a more benign and sustainable future. Topics like 'frugality' and 'mindfulness' have been explained in the context of sustainable behaviour thus, forcing the readers to think in terms of 'action'. Overall, a great contribution to this stream of knowledge, useful for behavioural scientists as well as climate change and sustainability experts."
—Shashi Kant, Director, *Master of Science in Sustainability Management Program, Institute for Management & Innovation, University of Toronto, Canada*

"The author dwells upon this book certain crucial issues pertaining to managing climate change and sustainability through myriad strategies of behavioural transformation. The book emphasizes that transformed human behaviour in a positive, responsible, ethical, accountable, and sagacious manner can effectively assist in providing feasible solutions to the problems of sustainable climate change. It also throws light on people's perceptions, innovative trans-disciplinary perspectives and the efficacy of systematic behavioural interventions. The reflections on psycho-spiritual and philosophical basis of sustainability make this book a unique contribution to human ecological analysis at the cross-national or cross-cultural levels."

—Prof. Ramesh K. Arora, *Chairman, Management Development Academy, India*

Contents

1 **Climate Change and Sustainability Behaviour Management**   1
   Introduction   1
   Climate Change and East–West Divide   3
   Sustainability   5
      What Is Sustainable Behaviour?   7
      A Long Way Between 'To Say' and 'To Do'   8
      Integrating Climate Protection with Sustainable Development Goals (SDGs)   10
      Sustainable Consumptive Behaviour   12
      Human–Climate Interface and Sustainability in Post COVID-19 Era   14
   Conclusion   20
   References   21

2 **Climate Change Risk Appraisal and Adaptation—Behavioural Processes**   25
   Introduction   25
   The Psychodynamics of Risk Behaviour   26
   Climate Risk Appraisal and Perception   31
   Determinants of Risk Perception   34
   Risk Perception Processes   36
   Barriers to Climate Risk Perception and Sustainability   39

	Climate Risk, Resilience, and Adaptation	40
	Conclusion	41
	References	42
3	**Climate Change and Sustainability Communication—A Socio-Psychological Analysis**	47
	Introduction	47
	Framework of Sustainability Communication	49
	Corporate Sustainability Communication	51
	Sustainable Consumption Communication—A Psychological Framework	61
	Climate Change Communication	63
	COVID-19, Climate Change, and Sustainability Communication	69
	Conclusion—Shaping of a Sustainable Society	72
	References	74
4	**Frugality and Innovation for Sustainability**	79
	Introduction	79
	The Concept of Frugality	81
	Frugality Across Diverse Societies and Cultures	82
	Frugality, Values, and Consumerism	84
	Frugal Traditions in Eastern Religions	86
	Simplistic Model of Bhutan—A Happy Little Kingdom	88
	Dayalbagh—An Indian Hermitage for Frugality	89
	Beyond Materialism—A Frugality Model	93
	Frugality-Based Innovations for Sustainability	95
	Psycho-Social Correlates of Frugality	99
	Conclusion	100
	References	101
5	**Integrating CSR with Climate Change and Sustainability**	105
	Introduction	105
	Levels of CSR—A Behavioural Analysis	106
	CSR—A Stakeholders' Analysis	108
	CSR and Sustainable Development Goals	112
	From Sustainable Development to Sustainability Behaviour	113

Social Process Reengineering (SPR) for Responsible Business
Practices 117
Challenges for CSR and Need for SPR 119
Integrating CSR with Environment and Climate Change 124
CSR in the Era of COVID-19—Psycho-Social Concerns 126
CSR for Pushing Digital Interface and Remote Learning 127
Integrating Health, Wellness, and Social Responsibility 128
COVID-19 and the CSR Transition 129
The Growing Optimism 130
Conclusion 131
References 132

6 **Behavioural Transformation for Sustainability
and Pro-Climate Action** 137
Introduction 137
The Concept of Behavioural Transformation 138
Behavioural Transformation Across Societies and Culture 141
An Interplay of Behavioural Dynamics 142
Behavioural Transformation for Sustainability 149
Sustainable Consumption Behaviour 150
Behavioural Transformation for Pro-Climate Action 154
Climate Change Mitigation Through Behavioural
Transformation 155
Barriers to Behavioural Transformation 159
Conclusion 161
References 162

7 **Contemplative Practices, Climate Change Adaptation,
and Sustainability Management** 169
Introduction 169
Perspectives of Climate Change and Sustainability 171
Psycho-Spiritual Basis of Sustainability 173
Role of Contemplative Practices 176
Spiritual Intelligence (SI) 177
Mindfulness 181
Co-influencing the Eastern and the Western Thoughts 192
Conclusion 193
References 194

8 Conclusion: Looking Through a Behavior-Centric
   Prism   199
   Reference   205

Glossary of Keywords   207

Index   211

Abbreviations

BBC	The British Broadcasting Corporation
BR	Behavioral Restraint
CC	Climate Change
CDM	Clean Development Mechanism
CEO	Chief Executive Officer
CO2	Carbon dioxide
COP-13	The 13th Conference of Parties
COVID	Coronavirus disease
CSR	Corporate Social Responsibility
DPE	Department of Public Enterprises
FSSD	Framework for Strategic Sustainable Development
GDP	Gross Domestic Product
GE	General Electric
GHG	Green House Gases
IAA	Intention, Attention and Action
ICT	Information and Communication Technology
ILO	International Labour Organization
IMF	International Monetary Fund
IQ	Intelligent Quotient
MI	Mindfulness
NASA	The National Aeronautics and Space Administration
NGO	Non-Governmental Organization
PET	Positive Existential Transcendence
PMT	Protection Motivation Theory
PPI	Positive Psychology Intervention
PSA	Positive Sustainability Actions

PSC	Positive Sustainability Cognitions
PSE	Positive Sustainability Emotions
RSP	Rooftop Solar Power
SDGs	Sustainable Development Goals
SI	Spiritual Intelligence
SOP	Standard Operating Procedures
SPR	Social Process Re-engineering
SRT	Social Representation Theory
UNDP	The United Nations Development Programme
UNESCO	United Nations Educational, Scientific and Cultural Organization
UNISDR	The United Nations International Strategy for Disaster Reduction
WCED	World Commission on Economic Development
WFP	World Food Programme
WWF	World Wide Fund

LIST OF FIGURES

Fig. 1.1	Integrated concerns for environmentally conscious behaviour	6
Fig. 1.2	Approaches to sustainability behaviour	6
Fig. 1.3	Behavioural dimensions of sustainability	7
Fig. 1.4	Consumer behaviour in relation to personality and motivation	13
Fig. 1.5	Impact of COVID-19 pandemic on human–climate interface and sustainability management	15
Fig. 2.1	Behavioural dimensions of human climate interface	29
Fig. 2.2	Sub-components of risk appraisal	32
Fig. 2.3	Progression of climate risk appraisal and adaptation	32
Fig. 2.4	Levels of operation of risk perception	34
Fig. 2.5	Processes guiding climate change risk perception	37
Fig. 3.1	Distinctive features of sustainability communication	50
Fig. 3.2	Sustainability communication framework	51
Fig. 3.3	Sustainability communication-internal and external	52
Fig. 3.4	Commitment vs. communication matrix	56
Fig. 3.5	A psychological framework of sustainability communication	62
Fig. 3.6	The process of climate change communication (Adapted from Fischhoff [1995])	65
Fig. 3.7	Arguments for/against fear appeal for C3	66
Fig. 4.1	Frugality—behavioural consumption vs. choice	83
Fig. 4.2	Characteristic features of frugal innovation	96
Fig. 5.1	Levels of CSR	107

Fig. 5.2	CSR for sustainability through stakeholders' integration (Based on Wipro's Sustainability report, 2010–11)	111
Fig. 5.3	Framework of corporate sustainable development	114
Fig. 5.4	Social process reengineering in CSR	118
Fig. 5.5	Execution of SPR in CSR context	119
Fig. 6.1	IAA framework for behavioural transformation	146
Fig. 6.2	Desirable behavioural actions for sustainable living	153
Fig. 6.3	Role of behavioural determinants in shaping specific behaviour	155
Fig. 6.4	4D barriers to climate action (based on Stoknes, 2015)	160
Fig. 7.1	Dimensions of global climate change and sustainability management	172
Fig. 7.2	Viewpoints of sustainability	172
Fig. 7.3	Dynamics of climate change—psycho-spiritual dimensions	173
Fig. 7.4	Progression of psycho-philosophical thoughts of sustainability	174
Fig. 7.5	Reflections of spiritual intelligence	179
Fig. 7.6	Spiritual intelligence for climate change and sustainability	180
Fig. 7.7	Framework of mindfulness for climate change/sustainability	184
Fig. 7.8	Intention-attention-attitude framework of mindfulness	185

CHAPTER 1

Climate Change and Sustainability Behaviour Management

INTRODUCTION

Climate change, as a global issue, might have differential effects on different countries, but influencing planet earth as a whole, thereby influencing the sustainability of resources. Current concern over global climate change stems, in part, from the evidence that its causes are anthropogenic in nature with its roots in human behaviour and intricately linked to sustainability management(Karl & Trenberth, 2003; Rosa et al., 2015). Hence, any change in the global environment warrants changes in human behaviour so that the sustainability of environment and resources can be better managed. Therefore, effective sustainability solutions must draw on a broader understanding of social systems and human behaviour (Ehrhardt-Martinez et al., 2009).

Much of the happenings around us are an outcome of psychological/behavioural encounters with our surroundings and external world or in other words, what one may call as 'human–climate interface'. This human–climate interface, including climate change, belief systems, and psychological responses, varies across societies and cultures. Therefore, people need to know what difference can be made by switching over from merely 'wanting change' to actually 'working for change' (Mudaliar & Rishi, 2012). Many times, global problems need a solution which mostly originates from the local roots. Therefore, multi-country research and

© The Author(s), under exclusive license to Springer Nature Singapore Pte Ltd. 2022
P. Rishi, *Managing Climate Change and Sustainability through Behavioural Transformation*, Sustainable Development Goals Series, https://doi.org/10.1007/978-981-16-8519-4_1

cooperation with inclusive approach towards development is the need of the hour to continue our battle against adversities of changing climate.

Climate change is seen as a slow and gradual process that has taken place over years and years. Though the impacts are very much obvious in recent times, still the required urgency is not being experienced by most of the people. The reason behind this is that gradual climatic changes do not elicit the required anxiety because of a feeling of lack of control over catastrophic consequences. Hence climate change is not perceived as an 'immediate risk'. For much of the general public, climate change is still an 'unknown risk' which is new and has unforeseeable consequences and therefore hard to believe or feel at risk about it. The greater the exposure to hazards, the more readily the adverse consequences are recognized by people. Public support or opposition to climate policies (e.g., treaties, regulations, taxes, and subsidies) will be greatly influenced by public perceptions of the risks and dangers posed by global climate change. Hence, effective solutions must draw on a broader understanding of social systems and human behaviour (Ehrhardt-Martinez, 2011).

Article 6 of UNFCCC also stresses for a country driven, cost-effective, and holistic approach following an interdisciplinary perspective. As a part of amended New Delhi Work Programme (2007) to implement Article 6 of the convention, six elements starting from education to training, public awareness, public access to information, public participation, and international cooperation were emphasized. Under public awareness, a special effort to foster psycho-behavioural change has been stressed (draft decision-cp.13/2007).[1]

Behavioural responses to climate distress and adaptation are essential concepts in the broader sphere of sustainability management. Although both awareness and insight in the importance of mitigation and adaptation have increased significantly in the last few decades; there is still a need to enhance the capacity of professionals in these vulnerable areas further to boost the behavioural adaptation process. The chapter will focus on the concept of climate change and sustainability management from a transdisciplinary perspective with a special emphasis on behavioural and socio-psychological approach of understanding them.

---

[1] https://unfccc.int/cop8/latest/14_cpl3_sbstal23add1.pdf.

CLIMATE CHANGE AND EAST–WEST DIVIDE

We are the first generation to feel the sting of climate change, and we are the last generation to do something about it.[2]
  Jay Inslee, Governor of Washington, 2019

In today's technologically progressing world, we are hastily moving towards urbanization and industrialization. At the same time, we are also experiencing the impacts of changing climate, primarily due to anthropogenic causes. However, globalization and urbanization are differentially instrumental in causing climatic adversities to different regions of planet earth and posing diversified challenges to sustainability management. Positioning sustainability in a system based perspective makes us understand the fact that focus on just specific elements may not be sufficient to understand the concept from a holistic perspective as no single discipline can account for the dynamics of sustainability behaviour and climate change.

Since time immemorial, development has always been a prerogative of the nations with abundance of money and power while the ones with limited resources have continued to bear the hardship of development. Though lagging far behind in the race of development, the eastern world would continue to strive hard in a rapid way for economic growth and a better quality of life. Hence, continues this divide of interests between the just and unjust ways of the eastern and the western world, respectively. In the never-ending debate of sustainability between the eastern and the western nations, those who are powerful, have not led but fled, as is evident from several experiences in the past where the western countries have shown reluctance to obey conventions and protocols that required them to reduce their carbon emissions. The way eastern world has behaved in their contributions to climate change and sustainability management is like a drop in the ocean when compared with the behaviour of their western counterparts.

Though fairness demands an initiative to be taken first by the western world to curb their emissions, yet, ironically, the eastern world is facing the dilemma of 'to do or not to do'. The focus is increasingly turning

---

[2] https://e360.yale.edu/features/tackling-climate-change-governor-jay-inslee-has-a-plan-for-that.

more towards the question of developing countries' emissions despite the fact that their per capita emissions on an average comes out to be merely one-sixth those of the industrialized nations (Chandler & Gwin, 2004). Sometimes it compels us to think that climate change is probably a myth created by western world to halt the developmental process of the east. However, ever-increasing heat and global warming along with increasing frequency of extreme climatic events, including floods, storms, cyclones, etc., restricts this stream of thought and forces both the east and the west to join hands in managing the changing climate, keeping a balanced and mutually beneficial perspective in mind.

Any step taken by developing countries under the pressure of superpowers to compromise with their ways of development in the name of climate change would mean entering compounded backwardness several times more than the existing situation. On the other hand, if they do not act the way they should and continue focusing upon building their economies in energy-costly ways, they are at increased risk of bearing the ills of global climate change. It is a real tough choice for developing nations in the east to arrive at any solution that enables them to protect the global environment without sacrificing their economic expansion activities. Howsoever tough the choice may be, a solution is required because in the event of climate change impacts, effective coping for the eastern developing nations would be even more tough. On one side they have higher concentration of population and on the other, the limited resources and technology support to address the issues of climatic adversities, which directly affect their marginalized population. Not to forget, increased incidences of extreme climatic events would be a direct hit on the eastern developing nations, exerting multiple strains on their already over-burdened economies.

Still, the mere acceptance of emission bindings cannot prove to be an exclusive decisive factor in believing that any particular region is behaving in an environmentally responsible way, targeting mitigation of climate change. There can be additional pro-environmental ways and efforts through which climate protection could be approached. Hence, the idea of developing technologically sound energy efficient means would emerge as an intelligent choice before those eastern developing nations who do not wish to call off their economic advancement.

Looking from the perspective of the western world, it is definitely their responsibility to act first but not at all their sole and absolute liability. Diverse risks have been administered to human health in the past by the skewed and convoluted nature of environmental and social exhibitions of climate change(McMichael, 2013). Sustainable development goals which are interrelated to each other are also bearing the consequences of the widespread disease. Sustainable planning of creating such ecosystems which reduce the transmission contact of pathogens among humans by national and international institutions to anticipate such unprecedented EID risks will be the next hour demand of achieving different SDGs (Di Marco et al., 2020).

Instead of lamenting upon who has done what in the past and arguing upon who is doing what in the present, the need of the hour is a shift of focus on an evenhanded responsibility sharing by both the developing and the developed countries, as ultimately they both belong to this planet earth in common. It is a fact, which cannot be neglected that climate change, largely, is a behavioural outcome of both eastern and western nations—a result of indifferent human behaviour towards mother earth. Behavioural blunders cannot be certainly reversed but repaired for sure. Both the developing and the developed nations need to join hands for this cause and planet preservation should be the biggest motivation for both east and the west, to act and introduce attitudinal changes irrespective of the division of interests and opinion between them.

SUSTAINABILITY

In today's globalizing world, technological innovations, carbon footprint and planet concerns go hand in hand leading towards a greener and environmentally conscious technological advancements as given in Fig. 1.1.

Sustainability is not only the question of resource use per person but it also talks about our ability to understand the science of sustainability, our ability to regulate our impact on environment, our beliefs and values for conservation and sustainability, our perceptions towards future climate change risks and our ability to negotiate climate change and sustainability solutions at both local and global level.

Sustainability can be interpreted from two perspectives. First is the traditional viewpoint in which sustainability concerns regarding ecological and social environment are discussed. In this perspective, sustainability focuses on avoiding the exploitation of resources, keeps a check on depletion of resources, and for their irreversible alteration. In this connection, it is known as a 'self-seeking ego-centric approach' (Fig. 1.2).

6 P. RISHI

Fig. 1.1 Integrated concerns for environmentally conscious behaviour

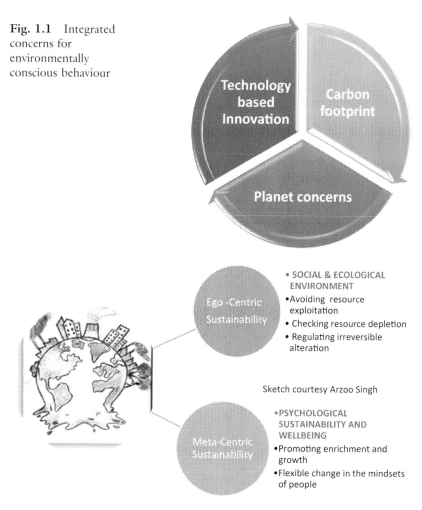

Fig. 1.2 Approaches to sustainability behaviour

The other view is known as modern viewpoint that promotes the psychological sustainability and well-being. In this perspective, sustainability focuses on promoting enrichment, growth, and flexible change in the mindsets of people and hence known as altruistic or meta-centric approach.

What Is Sustainable Behaviour?

Sustainable Behaviour encompasses within itself the awareness about responsible consumption and conservation of resources at cognitive level; human belief systems and values inclined towards responsible resource consumptive practices at affective level; and adopting sustainable practices in life at conative level (Fig. 1.3). It demonstrates responsible resource conservation behaviour at the micro and macro level with the intend to contribute towards well-being of present and future generations.

Anticipating behavioural change to promote sustainable resource management practices is one of the key agenda for environmentalists and social scientists. It can either be in the form of behavioural reduction like reducing practices which may cause harm to environment and society (varied forms of pollutions) or behavioural summation, to learn and practice behaviours which may add to healthy lifestyles of self to

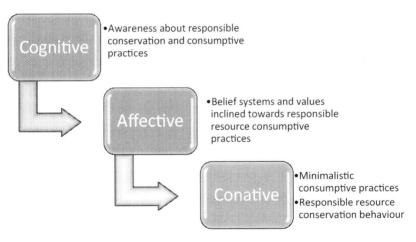

Fig. 1.3 Behavioural dimensions of sustainability

reduce individual and societal cost of health services (Collier et al., 2010). In short, it is underpinning the positive behaviour which is sustainable and in the desired direction, through incentives or normative pressure or feel-good factor and addressing the socially and ecologically detrimental behaviour by imposing disincentives, the details of which are given below:

A Long Way Between 'To Say' and 'To Do'

Communication about 'being sustainable' is having limited efficacy to convert into sustainable practices. Humans operate at a very limited level of rationality when the question of sacrificing their comfort zone, freedom to act, and privileges granted by virtue of their position of power are concerned. For example, we all are very comfortable preaching to public about contributions of air conditioning in CO_2 emissions and its resulting effect on global climate but not using or limiting its use in the official chamber is a highly challenging decision to take. Michelle Shiota remarks that explicit sustainability attitudes have limited predictive value in terms of behaviour(Shiota et al., 2021). Our habits, impulses, and desire for comfort and convenience have trouble competing with even our best intentions and dearly held beliefs. Hence, the statement by Unilever's Richard L Wright that successful communication requires a very high value of engagement[3] is very promising.

Context Above Content
Behavioural change is context specific and similarly, sustainability behaviour also operates in a context specific manner. India, including the eastern World, especially adopts the behavioural model of food sustainability which suits their traditions and culture, well-being and health, and also economical to procure. Children are advised not to waste the food keeping in mind the needs of millions of people who are deprived of required nutritional supplements due to limited resources. However, the same eastern world does not mind spending hard earned resources in abundance to feed thousands of people on the occasion of family wedding and do not even mind wasting food. According to them it suits to the context and not all the times, sustainability lessons, we preach, can be practised. Giving cognizance to that India's Union Ministry of Consumer

---

[3] https://www.theguardian.com/sustainable-business/behavioural-insights/behaviour-change-sustainability-debate.

affairs, food and public distribution took an initiative to regulate the food wastage in restaurants considering that it is a criminal waste of national resources which belong to humanity. So environment plays a key role in creating sustained behavioural change. Calr Hughes, from Wales Centre for Behavioural change, remarks *'We have to identify behaviour we would like to see, then arrange rewarding environments or disincentives for undesirable behaviour'*.[4]

In the western world, there are examples where food waste is punishable. Take the case of Germany with per capita income over $44100 as per IMF estimates. But wasting food is a punishable offence by many restaurants as resources belong to the country and society, which need to be used in a responsible manner. Similar policies are in place in many restaurants of France and Switzerland too. With changing climate, ensuring food sustainability is one of the biggest challenge. Therefore, in both eastern or western countries, wastage is completely undesirable to ensure food sustainability as it affects humanity across the globe.

The Power of Social Norms
Human beings have a general tendency to follow social norms. If most people are following a sustainability practice, rest feel motivated to do so as they want to go with the larger chunk of population in order to experience behavioural safety. Take the example of littering in public places. Even the habitually careless people will not litter at the places where no existing trash is lying, considering it as the social norms of that place. Similarly, comparing the eastern and western norms of honking behaviour, eastern people are accustomed to honk in their home place but will not honk in the western world, while driving, considering it as breaking the social norm. It is even more effective than placing a board of 'no honking please'. Mike Daniels states that concerning environmental sustainability social norms play a very powerful role in any behavioural change context.[5]

---

[4] https://www.habits.ninja/en/2016/11/05/about-behaviour-change-and-sustainability/.

[5] https://www.theguardian.com/sustainable-business/behavioural-insights/behaviour-change-sustainability-debate.

Repeated Reminders for Incremental Behavioural Change
There have been worldwide efforts to reduce single use plastic and make carry bag chargeable in shopping malls and other places. It is intended to push the change for bringing your own bag and its reuse keeping the environmental impact at the forefront and reminds at every checkout from the store when you pay for bag and more than that, when you see others carrying their own bags. These repeated mental reminders strengthen the desirable and environmentally friendly behaviour in an incremental manner and over the period, behaviour is likely to be established.

Triggering a Value Change
Short-term behavioural change is quickly possible with incentives and extrinsic motivation. However, such behavioural change is not sustainable over the period as it does not trigger value change. Once, the incentive is withdrawn, the rebound effect of behaviour is likely to take place. Intrinsic behaviour is value based which changes the behaviour from within, irrespective of external reward or incentive. Such deep-seated behavioural changes are likely to be more sustainable over time in view of their established connection with our value system. Therefore, there is a need to move from 'wanting change' to actually 'working for change' adopting various modes of behavioural change process as described in subsequent chapters.

Integrating Climate Protection with Sustainable Development Goals (SDGs)

The need for economic growth in developing countries must be met through ways and strategies that do not conflict with climate change and sustainability objectives. Pandemic has created threats as well as opportunities for both eastern and the western world, for sustainable development by creating systemic changes in both lifestyle and behaviour. There have been instances where air pollution and tobacco smoking have caused higher COVID-19 mortality rates. Similarly, deforestation can increase wildlife-originated pathogens' exposure to humans and climate change induced melting of ice has the potential to release undiscovered viruses, which might have been lying dormant under the frozen ice sheets.[6]

---

[6] https://www.livemint.com/opinion/online-views/opinion-india-s-opportunity-for-sustainable-growth-11587716965607.html.

Hence, there is a need to re-establish the link between environment, people, health, and economy through promotion of sustainable infrastructure with a special focus on health support systems. Besides, eastern countries with the abundance of sunshine are expected to support renewable energy systems, especially rooftop solar panels, to support critical health services in remote areas where power constraints are visibly high. Scaling up of electrification and investment in cold storage facilities in remote areas can also be largely beneficial to ensure food availability in a sustainable manner, even though there are limitations in transportation. Its recent example is Dayalbagh Educational Institute of India which is demonstrating 'low-cost-high-efficiency techno-smart models to provide quality food, air, water, energy and health care services' by creating 'a holistic and sustainable habitat'..... 'Simple living, in harmony with nature and practicing economy in all activities, smart use of technology, minimalistic approach in all aspects of life, implementation of the Jugaad concept, i.e., innovative, flexible and costeffective solutions are some significant features of the Dayalbagh way of life'[7] contributing to sustainability behaviour as integrated in their educational policy.

Pursuance of Sustainable Development Strategies can help mitigation of Climate Change.[8] It is possible through the adoption of energy-efficient electricity generation and transmission; moving towards renewable sources of energy for sustainable supply; effective public transport systems; adopting sustainable and participatory forest management practices to conserve biodiversity; protect watersheds; promote rural employment; and enhance income generation of forest resource-dependent communities which will, in turn, reduce pollution, greenhouse gas emissions and carbon sequestration.

Integrated development strategies, concurrently taking care of climate protection objectives, can be subservient in achieving sustainable development targets in the eastern world. International community must appreciate that its efforts to address climate change and the needs of countries with low developmental indices are not compromised. Equitably speaking, the vulnerable countries are primarily the ones with minimum

---

[7] https://www.dei.ac.in/dei/edei/files/ASSESSMENT%20OF%20THE%20BEST%20PRACTICES%20OF%20DAYALBAGH%20EDUCATIONAL%20INSTITUTE%20TO%20COMBAT%C2%A0%C2%A0THE%20COVID-19%20CHALLENGE.pdf.

[8] Climate change, sustainable development and India: Global and national concerns: Jayant Sathaye, P. R. Shukla and N. H. Ravindranath (reproduced with permission).

contributions to the current GHG emissions.[9] Hence, the normal liability lies with the economically advanced countries to lend a helping hand in the achievement of sustainable development goals in the manner through which climate change adaptation and mitigation objectives are duly taken care of.

Sustainable Consumptive Behaviour

Mahatma Gandhi *rightly stated that 'The world has enough for everyone's needs, but not everyone's greed'*.[10] World has no dearth of resources to feed the planet but the manner in which these resources are being relentlessly subjugated by humans will cause rapid increase in our material footprint, overtaking the population growth and corresponding developmental needs. Hence, the anticipated increase in global population to nearly 9.6 billion by 2050, will necessitate three planets to meet the natural resources requirement in order to ensure the sustenance of current lifestyles.

Therefore, a sustainable future is critically linked to the way we systematically proceed towards sustainable consumptive behaviour. Sustainable Consumption and Production is one of the Sustainable Development Goals (SDGs) of United Nations. It is interpreted as utilization of products and *services related to basic living and sustaining requirements to ensure a better quality of life. At the same time, minimalistic approach towards exploitation of natural resources, and release of toxic materials, waste and pollutants over the life cycle of product or service in order to safeguard the requirements of generations to come.* Unsustainable consumptive behaviour is not only an ecological challenge but a social and financial challenge too. Tim Jackson, aptly remarks in his book, Prosperity without Growth *'People are persuaded to spend money we don't have, on things we don't need, to create impressions that won't last, on people we don't care about'*.[11]

---

[9] Baumert et al. (2005) estimate cumulative historic contributions (1850–2002) to climate change at over 75% for developed countries, with the United States and EU-25 alone constituting 56%.

[10] https://www.goodreads.com/quotes/427443-the-world-has-enough-for-everyone-s-need-but-not-enough.

[11] https://www.goodreads.com/author/quotes/177609.Tim_Jackson.

Consumer behaviour operates on the principles of behavioural science where 'motivation to buy' plays a very important role in what we finally decide to buy. Human motivation is closely linked to personality and the way we manage our social relationships. Human beings are broadly classified as introverts, ambiverts, and extroverts and at the same time, have intrinsic and extrinsic motivations. Intrinsically motivated behaviour is self-guided with no or minimal intent to get approval/appreciation of others including incentives or rewards. However, the reverse is true for extrinsic motivation, which is dependent on external reinforcement/incentives, to execute any sustainable consumer behaviour. People who are extroverts with high degree of sociability and need for external approval, are likely to look for extrinsic motivations and their consumptive behaviour is guided by what others are doing in the social strata they operate (Fig. 1.4).

Hence, they are less likely to apply their own logic of moving towards sustainable consumptive behaviour, if others around them are not doing so. However, it is a very important area to conduct empirical research, in order to promote sustainable consumer behaviour. It can have its larger impact on the planet in terms of achieving SDG as well as addressing the adversities of changing climate.

Pandemic COVID-19 is also likely to promote sustainable consumptive practices as in view of financial constraints on people, consumption

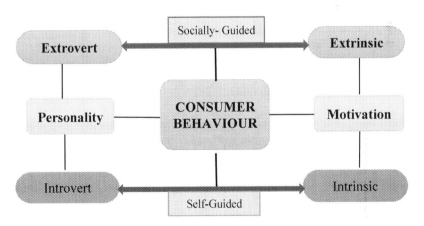

Fig. 1.4 Consumer behaviour in relation to personality and motivation

of non-essential goods is being thoroughly reassessed by people across all the countries. Besides, in the times of social distancing and minimum interactions with people, giving more opportunities for self-introspection and family bonding, there has been a change in the perspective towards living life. It may also influence the way people consume things, which was earlier influenced by social media standpoint of 'perfect life and looks', advertising and other extrinsic factors. There is a SHIFT framework which includes five broad psychological routes to encourage sustainable consumer behaviour change: Social influence, Habit formation, the Individual self, Feelings and cognition, and Tangibility which will be helpful in fostering sustainable consumer behaviour (White et al., 2019).

A study carried out by Accenture in early April 2020 through a survey of fifteen countries showed that consumptive practices are getting more sustainable across the globe and the likelihood of these changes to stay further in post-pandemic era is high. Survey data shows that 60% of respondents were spending more time on personal care and psychological well-being along with more focus on physical fitness. Further, 64% of consumers were concerned about limiting food waste consistently while 50% of consumers were consciously buying healthy products and were likely to do so in future too. Besides, 45% of consumers were making sustainable choices while buying products and services, moving a step ahead towards SDG—Sustainable products and services.[12]

Hence, it is a valid research proposition to explore the impact of pandemic on the promotion of intrinsic motivation in regard to consumptive practices among the diverse population of east and west and its impact on sustainable consumption of products and services.

Human–Climate Interface and Sustainability in Post COVID-19 Era

COVID-19 pandemic has created global havoc and health crisis forcing slowdown of economies within strict lockdown measures. It resulted in reduced commuting leaving an impact on the environment and climate change in terms of reducing CO_2 emissions, and on the sustainability of resources. It is an interesting issue to ponder upon as described in Fig. 1.5. During this phase, the human–climate interface and sustainability issues are likely to be projected in a substantially different manner.

[12] https://www.accenture.com/_acnmedia/PDF-130/Accenture-Retail-Research-POV-Wave-Seven.pdf.

Fig. 1.5 Impact of COVID 19 pandemic on human–climate interface and sustainability management

It will be in one way, to help address the climate change and associated adversities and in another way to challenge our already undergoing efforts towards carbon–neutral initiatives and resource conservation. Continuance of such impacts over a longer period can positively as well as negatively influence our efforts to address the issues of climate change and sustainability management in east and west.

Climate analysts say that global emission of CO_2 has been reduced by 0.3% because of confinement measures for transportation and other environment degradation sectors during suspended time of the crisis. According to some, the pandemic has happened to be a 'blessing on disguise' bringing air quality improvement for the next generation but nobody wanted it in this way by adversely affecting jobs, mental lives, and creation of global health crisis (Tahir & Batool, 2020).

Exponential Overload of Human–ICT Interface

Each one of us is facing an exponential overload of human–technology interface. At one side, we observe the rapidly emerging digital technologies in almost every front, whether academic or financial or governance or health management, to match the needs of constrained mobility and physical distancing norms of society. At another side, we observe strained minds and bodies struggling to match the ever-increasing expectations from them at work and all other day-to-day spheres of life. While undergoing ICT-interface, there are large-scale bandwidth issues, created due to exponential increase in network demand, especially in the east, with vast diversity of resource availability and technology acceptance.

Behavioural factors like readiness to accept ICT breakthrough vs. technological aversion and forced compliance are the likely decisive factors in how differential approaches to cope with digital stress are adopted (CDP & IIM Bangalore, 2014).

Limited Mobility and Need for Transportation
The large-scale mobility of people at national as well as international level is an emerging challenge for humanity and natural environment (Eshraghian et al., 2020). Pandemic has imposed severe constraints on human mobility in view of the limitations of public transportation, standard operating procedures for distancing, and quarantine guidelines. Non-essential travel such as ecotourism and international leisure visits is likely to remain suspended further for months (Bakar & Rosbi, 2020). This imposed curtailment of human movement may result in alterations in human mobility patterns across the society (Chakraborty & Maity, 2020). At the same time, it has opened a wide array of safe and sustainable options so that essentials of working are not disregarded. The surge of ICT solutions in diversified domains during pandemic phase is enabling sustainable management of resources to a large extent, especially by providing options to physical mobility through range of safe and secure online meeting rooms reducing the need for actual movement and transportation. After being habituated with online meetings and conferences, it is likely to have long time positive impact on reducing GHG emissions and resulting pollution levels in almost all parts of the world, which can be a very significant development to move towards a carbon–neutral society and a positive step towards Sustainable Development Goals (SDGs). Research studies have recorded that proper nationwide lockdown strategies can pave the ways for reduction of approximately 1000 Mt GHG in next five years (Kumar et al., 2020).

Constrained Human–Human Interface and Social Events
At the behavioural front, there has been a major setback for people, especially the extroverts who have been forced to limit their interface with other people due to substantial restrictions in organizing social events. ICT can substitute work-related human–human interface but the real feel of human touch in relationships and social as well as emotional connectivity in events is irreplaceable through electronic modes. However, these constraints, for a shorter period of time, convey no long-term harms to socio-emotional relationship building and largely contribute positively

to sustainable development. Reason being that it broadly cuts short the need for non-essential transportation for business/academic meetings and events, which can be efficiently managed through virtual modes with minimum wastage of time and contributing to reduce transportation-linked air pollution and global warming. Hence, keeping virtual modes of interactions, especially for professional communications and events, should always remain as an efficient alternative for long-term promotion of sustainability.

Regulated Industrial and Construction Activities
Industrialization, an important process for economic growth and development of society and environmental conservation have always remained at logger's head with each other for centuries, whether in east or in the west. There has been imposition of large-scale regulations in industrial production and other activities including construction when nationwide lockdowns were imposed in various countries across the globe to contain the pandemic. As a result, a significant reduction in pollutants across many cities of India, especially in Delhi Mega city, amidst the lockdown period was found, forcing people to think how regulated industrial activities and transportation can be adopted as effective strategies to control air pollution (Mahato et al., 2020). The nature of the current pandemic on emissions and climate change points to the challenge that the sustainable way to achieve long-term CO_2 emissions can be led majorly with the support of technology that has adopted the decarbonizing factors. As economic growth has slowed down, emissions have also fallen down substantially in countries like United States, China, United Kingdom, India and also in rest of the world and industry carbon emission fallout is probably its major reason (Hausfather, 2020).

Changing Need Cycle and Consumptive Patterns
In the post-COVID era, people are likely to experience significant changes in their need cycle which are going to be more rational and logical rather than impulsive. There has been a significant drop in the consumption levels of non-essential products and unprecedented shift in consumer preferences primarily due to their closure as well as financial concerns.[13] More

---

[13] https://www.lexology.com/library/detail.aspx?g=877406e6-5365-40a2-9d20-3e5 2a3fc6522.

so, a major shift towards online shopping due to hygiene and contagiousness concerns has also been observed. Besides, people are likely to spend more on groceries, household products, personal transport, and digital needs which is going to be the new normal. McKinsey market research says that there is a significant decline in consumer optimism in most of the countries during pandemic with China and India still maintaining their optimism levels to shop. However, Japan is following a conservative approach and most of the European countries are expressing substantial pessimism in their consumptive patterns. 70% of consumers suffering from pandemic outbreak are anticipating major re-adjustments due to the work-from-home practices.[14]

Social identity and sustainable behaviour of humans have been a powerful influence on the actions related to climate change as well as the environment. If a group or a society has a pro-environmental approach in its decisions and actions, then the whole group will develop a similar identity in their behaviour whereas intergroup comparisons and negative intergroup relations may cause conflict and hinder the pro-environment-based consensus of the more sustainable world. Strategies to reduce the intergroup environmental conflicts and understanding the influence of behavioural norms on climate change will help to promote a more environmentally sustainable world (Fielding & Hornsey, 2016).

Although behavioural changes are an ongoing process but the entire world has been forced into habitual transformation and if we go with the old saying that '21 days to form a habit' the altered life style has continued much beyond that. Therefore, with gradual unlocking of economy and the newly formed consumptive patterns are likely to stick around for a length of time. However, how these transformed consumption patterns are going to affect the sustainability agenda and business world, is a big question to ponder.

Controlled Interference with Nature and Biodiversity
In order to contain the COVID-19 pandemic spread, two-thirds of the world's population was under lockdown and 4.5 billion humans were confined. This was viewed by Bates et al. (2020) as an unprecedented largest human lockdown which can be termed as a Global Human

---

[14] https://www.mckinsey.com/business-functions/marketing-and-sales/our-insights/a-global-view-of-how-consumer-behavior-is-changing-amid-covid-19#.

Confinement Experiment.[15] It has provided a distinctive opportunity to scientists to explore the impact of human actions, especially the mobility, on a wide array of ecosystems, wildlife, protected areas, and the processes regulating biodiversity. There is a need for natural resource managers and environmental scientists, to share their country specific observations so that a comprehensive global understanding can be built upon and further exploratory studies can be conducted. Many reports reviewed that environmental degradation is not permanent and can prove to be reversible provided with a recovery time as witnessed in the initial months of the suspension of activities during pandemic. During these times, policies can be reconsidered with a new dimension and it has been proven that environmental conditions can be mitigated by planned strategic measures (Lokhandwala & Gautam, 2020). In the long term, pandemic will offer lessons and opportunities leading to possible climate/sustainability action. The collective global observations can bring out unanticipated insights to take further steps regarding conservation strategies for ecosystem and biodiversity.

Outsourcing of Services from West to East
For years now, companies belonging to developed nations in the west are outsourcing major aspects of their businesses to the developing nations in the East. While this leads to cost efficiency and better profits for the host companies, it also helps in providing jobs and requisite economic push to the developing nations. Low wages, less stringent labour laws, and abundance of resources make east a go-to destination for outsourcing. However, there is a need to ponder upon the ecological impact of this practice. Peters and Janssens-Maenhout (2012) reported that carbon emissions have steadily declined in the Organisation for Economic Co-operation and Development (OECD) nations while there has been a surge in carbon emissions in the developing nations of the East. Shifting of manufacturing bases from the West to East has played an important role in these trends. OECD countries are managing their emission by outsourcing to comply with the Paris climate deal, but the quest for economic growth in the East has led to higher carbon emissions. With the arrival of foreign companies, western culture has also influenced the developing countries in the east leading to development of new habits

---

[15] https://www.sciencedirect.com/science/article/pii/S0006320720307230.

and trends. These new trends have led to increase in consumerism while putting pressure on sparsely available local resources.

Conclusion

Rapid industrialization and urbanization in the recent past had an adverse impact on our planet. Most of the developmental activities are being led by the rich and powerful western countries while the developing countries in the east are putting in strides to cope up with the west. Reluctance of the western countries in following the climate policies clearly shows that they have missed the chance to lead the fight on climate change. However, a solution needs to be formulated as most of the eastern countries lack the technology to curb the adverse effects of climate change especially on their marginalized communities. Sustainable behaviour anticipates responsibly reducing carbon emissions and having environmental concerns for future generations as well. To make it happen, behavioural changes play a pertinent role. COVID-19 Pandemic has steered some significant changes in consumptive patterns leading to a rational shift in consumer's preference to spend on more day-to-day needs.

Dual perspectives of sustainability mentioned as Ego-Centric Sustainability and Meta-Centric Sustainability are discussed. It is high time to move from preaching sustainability to practising sustainability. Social norms play an important role in promoting sustainable behavioural change. Though, short-term behavioural changes can be easily achieved, over all, in a long run, value based behaviour changes prove to be more useful. Effective adoption of Sustainable Development Goals can help in mitigation of climate change through attitudinal changes in favour of sustainable behaviours, incorporating cognitive, conative and affective elements. While extroverts are less likely to apply their own logic in moving towards sustainable consumption, introverts are more likely to take self-guided route to the same.

Studies and scientific literature have reviewed that social vulnerability is also one of the dimension which is impacted by climate change affecting a wide social and demographic group. This wider section of society is majorly constituted of the low and the middle level income groups where there is unequal distribution of the required economic resources and social safety nets. The impact on certain human well-being dimensions such as health, safety, food security, and displacement drastically increases after certain thresholds in social vulnerability are crossed by the physical climatic variability (Otto et al., 2017). Analysis of risk appraisal and perception in regard to climate change also plays an important role in this regard as elaborated in subsequent chapters..

REFERENCES

Bakar, N. A., & Rosbi, S. (2020). Effect of coronavirus disease (COVID-19) to tourism industry. *International Journal of Advanced Engineering Research and Science*, 7(4).

Bates, A. E., Primack, R. B., Moraga, P., & Duarte, C. M. (2020). COVID-19 pandemic and associated lockdown as a "Global Human Confinement Experiment" to investigate biodiversity conservation. *Biological Conservation*, 108665.

Baumert, K., et al. (2005). *Navigating the numbers: Greenhouse gas data and international climate policy*. Washington, World Resources Institute

CDP, & IIM Bangalore. (2014). *ICT sectors role in climate change mitigation*. https://6fefcbb86e61af1b2fc4-c70d8ead6ced550b4d987d7c03fcdd1d.ssl.cf3.rackcdn.com/cms/reports/documents/000/000/860/original/CDP-ICT-sector-report-2014.PDF?1472041398

Chakraborty, I., & Maity, P. (2020). COVID-19 outbreak: Migration, effects on society, global environment and prevention. *Science of the Total Environment*, 138882.

Chandler, W., & Gwin, H. (2004). China's energy and emissions: A turning point. *Data Set*. Available Online. http://www.Pnl.Gov/Aisu/Pubs/Chandgwin.pdf from Pacific Northwest National Laboratory, Richland, Washington.

Collier, A., Cotterill, A., Everett, T., Muckle, R., Pike, T., & Vanstone, A. (2010). *Understanding and influencing behaviours: A review of social research, economics and policy making in Defra*. DEFRA.

Di Marco, M., Baker, M. L., Daszak, P., De Barro, P., Eskew, E. A., Godde, C. M., Harwood, T. D., Herrero, M., Hoskins, A. J., & Johnson, E. (2020). Opinion: Sustainable development must account for pandemic risk. *Proceedings of the National Academy of Sciences*, 117(8), 3888–3892.

Ehrhardt-Martinez, K. (2011). Changing habits, lifestyles and choices: The behaviours that drive feedback-induced energy savings. *Proceedings of the 2011 ECEEE Summer Study on Energy Efficiency in Buildings, Toulon, France, 2011*, 6–11.

Ehrhardt-Martinez, K., Laitner, J. A., & Keating, K. M. (2009). *Pursuing energy-efficient behavior in a regulatory environment: Motivating policymakers, program administrators, and program implementers*.

Eshraghian, E. A., Ferdos, S. N., & Mehta, S. R. (2020). The impact of human mobility on regional and global efforts to control HIV transmission. *Viruses, 12*(1), 67.
Fielding, K. S., & Hornsey, M. J. (2016). A social identity analysis of climate change and environmental attitudes and behaviors: Insights and opportunities. *Frontiers in Psychology, 7,* 121.
Hausfather, Z. (2020). *COVID-19 could result in much larger CO_2 drop in 2020.* The Breakthrough Institute. https://thebreakthrough.org/issues/energy/covid-co2-drop
Karl, T. R., & Trenberth, K. E. (2003). Modern global climate change. *Science, 302*(5651), 1719–1723.
Kumar, S., Bhardwaj, S., Singh, A., Singh, H. K., Singh, P., & Sharma, U. K. (2020). *Environmental impact of Corona virus (COVID-19) and nationwide lockdown in India: An alarm to future lockdown strategies.*
Lokhandwala, S., & Gautam, P. (2020). Indirect impact of COVID-19 on environment: A brief study in Indian context. *Environmental Research,* 109807.
Mahato, S., Pal, S., & Ghosh, K. G. (2020). Effect of lockdown amid COVID-19 pandemic on air quality of the megacity Delhi, India. *Science of the Total Environment,* 139086.
McMichael, A. J. (2013). Globalization, climate change, and human health. *New England Journal of Medicine, 368*(14), 1335–1343.
Mudaliar, R., & Rishi, P. (2012). A psychological perspective on climate stress in coastal India. In J. Scheffran, M. Brzoska, H. G. Brauch, P. M. Link, & J. Schilling (Eds.), *Climate change, human security and violent conflict: challenges for societal stability* (pp. 613–631). Springer. https://doi.org/10.1007/978-3-642-28626-1_30
Otto, I. M., Reckien, D., Reyer, C. P., Marcus, R., Le Masson, V., Jones, L., Norton, A., & Serdeczny, O. (2017). Social vulnerability to climate change: A review of concepts and evidence. *Regional Environmental Change, 17*(6), 1651–1662.
Peters, J. A., & Janssens-Maenhout, G. (2012). *Trends in global CO_2 emissions 2012 report.*
Rosa, E. A., Rudel, T. K., York, R., Jorgenson, A. K., & Dietz, T. (2015). The human (anthropogenic) driving forces of global climate change. *Climate Change and Society: Sociological Perspectives, 2,* 32–60.
Sathaye, J., Shukla, P. R., & Ravindranath, N. H. (2006). Climate change, sustainable development and India: Global and national concerns. *Current Science, 90*(3), 314–325. http://www.jstor.org/stable/24091865
Shiota, M. N., Papies, E. K., Preston, S. D., & Sauter, D. A. (2021). Positive affect and behavior change. *Current Opinion in Behavioral Sciences, 39,* 222–228. https://doi.org/10.1016/j.cobeha.2021.04.022

Tahir, M. B., & Batool, A. (2020). COVID-19: Healthy environmental impact for public safety and menaces oil market. *Science of the Total Environment*, 140054.

White, K., Habib, R., & Hardisty, D. J. (2019). How to SHIFT consumer behaviors to be more sustainable: A literature review and guiding framework. *Journal of Marketing*, *83*(3), 22–49.

CHAPTER 2

Climate Change Risk Appraisal and Adaptation—Behavioural Processes

INTRODUCTION

Much of the happenings around us are an aftermath of behavioural encounters with our natural surroundings to what we may call as 'human–climate interface'. Besides natural reasons, climate change has occurred majorly because of human callousness or apathetic behaviour towards our environment. Stott and Sullivan (2000) remarks that 'The Science of environment is socially and politically situated besides the subjective location of human perception'. Presently, a major chunk of information and scientific data on climate change pertains to, societal contexts and are interlinked to the vulnerabilities and social volatility of diverse segments of continents across the world. Given this significance, the reaction of societies to the problem remains sluggish and inefficient, seeking more input on this viewpoint (Hackmann et al., 2014). While climate change continues to intensify at a rapid pace, strategies need to be advanced that focus not only on the biophysical component of the issue, but also on the social, political, economic, and cultural complexities emphasizing on people's attitudes and values, their behaviours, practices, and the governance system (Weaver et al., 2014). It recommends integration of social and biophysical sciences to address the risk of climate change and fosters a better hold of the root mechanisms of climate change, including the behaviours and interactions of individuals, societies, markets, nations and other organizations, as well as mechanisms that turn information into practice.

Current concern over global climate change stems, in part, from the predominant evidence that its causes are anthropogenic: the result of human behaviour. Effective solutions must draw on a broader understanding of social systems and human behaviour (Ehrhardt-Martinez, 2011).

Review of existing literature supports the evidence that climate change is caused due to human interference with nature, natural resources, and their living style, hence it is important to know what humans think of, feel about, and wish to do about this problem. Cognitive analysis assesses the knowledge base and awareness level of respondents about various dimensions of climate change. It also includes people's perceptions and basic factual knowledge about climate change issues as being understood by them. However, just having awareness and risk perception is not enough as one has to resort to '*Aware but beware*' approach, to move ahead on the path of '*Climate walk*' besides merely being involved into '*Climate talk*'. For example, leaders are still debating on the what-when situation of commitment despite several years have been passed since Paris agreement has been approved. It is time to walk the climate talk and take substantial action.[1] The chapter will focus on how different models of risk appraisal can be globally adapted to facilitate behavioural adaptation with climate change in diverse global perspectives.

THE PSYCHODYNAMICS OF RISK BEHAVIOUR

When we think about some of the psychological issues engulfing the phenomenon of climate change, the scenario may appear like this. Phenomenon is global but our perspective is mostly local. Climate change happens in large time span however, our thinking is primarily limited to shorter time spans. It is a complex phenomenon with lot of uncertainties but we want to perceive things which are certain and simple to resolve. Therefore, the worry is not changing climate, the worry is our perceptual barriers and resulting inaction.

Behavioural science has a lot to contribute to our understanding of human behaviour in different realms including climate change to bring us

[1] https://www.signify.com/global/our-company/blog/sustainability/walk-the-climate-talk.

out of the denial mode of climate change like 'it is not happening' or 'it is natural' or 'it is human originated but others, not me, are responsible' or 'it is too late. What can we do now'? and so on and so forth.

Psychological interventions in the domain of climate change are the need of the hour in view of the increasing importance of behavioural adaptation and coping in climate change literature. 'Why do we behave the way we behave?' is the focus of attention for psychologists in differential contexts. They are constantly exploring the reasons behind this human indifference across the globe. Behavioural scientists, for many decades by now, have dedicated inclination to explore and analyse how common persons perceive climate change and other associated risks via behaviour impact cycles.[2] It consists of an interplay of feelings, emotions, experiences, and memories with a fundamental premise that the way one behaves can have a cyclic impact on his own behaviour as well as others. So if we are consistently behaving in a way that is climatically desirable and ecologically sustainable, it can have an impact on our future behaviour as well as that of others.

Climate change awareness is defined by UNISDR (2009) as 'the process of informing the general population, increasing levels of consciousness about risks and how people can act to reduce their exposure to hazards'.[3] Iturriza et al. (2020), further explained awareness behaviour as the interaction between experience, attention, and knowledge. Appraisal of risk and coping potential are significant predictors of behaviour intended to prevent any climate change linked to extreme weather events like cloud bursts, floods, hurricanes, snow storms, draughts, heat waves, etc. To mitigate climate change or to cope with its real or potential consequences, people must perceive it as a risk that they can or have encountered (risk appraisal).

The theoretical framework, Protection Motivation Theory (PMT) (Rogers, 1975, 1983) is considered as a conspicuous framework to explain the behavioural intentions to threatening environmental events. Core psychological research is dedicated on the underlying cognitive processes which people come across while confronted with risks. Psychological

---

[2] https://blue-sky.co.uk/tool/behaviour-impact-cycle-understanding-why-we-behave-as-we-do/.

[3] http://www.unisdr.org/we/inform/terminology.

research post-1960s conceptualizes risk perception as the social representation (Moscovici & Lage, 1976) of risk with a major prominence to factors beyond individual information processing (Joffe & Rosenbaum, 1999). Social Representation Theory (SRT) integrates both the process of elaborating representations and the emerging structures of thought (Duveen, 2000).

Noteworthy psychological variables in this regard include cognitive, affective, and conative aspects, which constitute the behavioural framework to explain climate change. Cop-13 amended New Delhi work programme,[4] in its implementation section, has indicated the need to conduct *knowledge–attitude–behaviours* surveys to establish a baseline of public awareness which can serve as a basis for further work and support the monitoring of the impact of activities.

In climate change context, the term *cognitive* refers to the higher level of mental process including aspects such as awareness, perception, reasoning, and judgement regarding the phenomenon of climate change, its dynamics and how it differentially impacts different regions across the globe. Varying levels of understanding at cognitive front plays a decisive role in the way people behave responsibly or otherwise, regarding climate change issues (van der Linden, 2015). Cognitive variables such as knowledge and awareness about climate change are prerequisites to make individuals perceptive about the risk associated with climate change. These psychological components are important for assessing risks and problems for sustainable development (Takala, 1991).

Iturriza et al. (2020) state that developing awareness about climate change is key to climate change resilience-building process, as it emboldens partnership and behavioural transformation. Following the triangulation approach through a thorough literature review, semi-structured interviews and case study, a framework for developing awareness in urban areas was developed with the incorporation of experience, attention, and knowledge, as few of the major constituents.

Only after knowing how informed the general population is about climate change, an inference can be drawn regarding the affective dimension of human psyche, associated to feelings and emotions. Awareness about climatic variability and its impacts may or may not generate affective concern among the masses about climate change. Slovic (2000),

[4] https://unfccc.int/cop8/latest/14_cpl3_sbstal23add1.pdf.

states 'it is tempting to conclude …that lay people's perceptions of risk are derived from emotion rather than reason, and hence should not be respected…'. Hence, in spite of the importance of affective dimensions of risk perception, it disseminates the relegation of its emotional perspective of common thought processes by favouring the logical and rational structures of the mind, i.e., rational information processing.

In a study of climate change perceptions in Britain and Italy, the researchers concluded that people are less willing to translate their concern about climate change at affective level into personal action because they perceive 'unwillingness of others to take action because of ignorance or lack of concern and the failure of institutions to provide leadership and effective legislation' (Biermann, 2002).

On the other hand, recent reference by (Shiota et al., 2021) again highlights the potential of emotions and affect in promoting desirable behavioural change in wider fields due to their motivational properties.

Authors further stress the pertinence of 'positive affect and emotions' over 'fear appeals' to promote climatically desirable/sustainable behaviour.

Next comes the *conative* aspect which pertains to the aspects of mental processes which are directed towards action or behavioural change. It plays a vital role to find out people's orientation to act and adapt their behaviour regarding climate change. A comprehensive analysis taking together the above three psychological variables in an integrated manner is expected to give promising solutions to psychologists studying climate change.

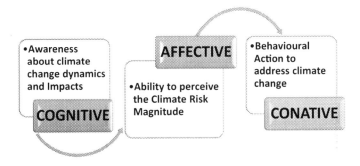

Fig. 2.1 Behavioural dimensions of human climate interface

For example, the western world, having an edge in terms of developmental indices and educational standards have adequate understanding about technicalities of climate change and how it is going to adversely impact the world. But at the same time, if they are not able to assess the magnitude of risk and are not having strong feelings and concerns regarding people (especially those who are going to suffer in different regions of the world, taking into considerations their economic limitations and low developmental indices), their personal commitment to change the lifestyle and adaptation efforts at micro as well as macro level will be limited. On the other hand, if the eastern world is not able to understand and appreciate the challenges, which climate change can pose in the times to come, the likelihood of any climate-conscious action/behaviour will be constrained in the absence of knowledge.

Hence, climate risk perception is challenging in both eastern and western worlds due to the common tendency of people to defend carbon intensive capitalistic lifestyle in order to maintain their comfort zone. People have a general predisposition that, 'Why should I contribute responsible behaviour towards climate change issues when others are not doing so?' Besides, they also have a belief that technology can provide them salvation from climate change linked adversities. So, whenever any sort of real or perceived inequality or inequity exists, cooperation tends to decline and people do not mind addressing the economic cost of climate change, if it doesn't come out of their own pockets.

In a study by Rishi and Mudaliar (2014), it was found that people from coastal zones of India were emotionally concerned about climate stress to a significant level and accepted that even though western countries are more accountable for climate change because of their developmental edge and lifestyle, responsibility also substantially lies on eastern nations too to adapt their lifestyles as they are going to be most adversely affected by climate stress because of their natural resource dependence and population size. Hence, integration of these behavioural components is desirable through coordinated efforts of eastern and western world through micro/macro level adaptation efforts and by negotiating solutions at local as well as global front to save the planet from climatic emergencies.

Psychodynamic extension of Social Representation Theory (SRT) postulated that, people when confronted with potential threat, express anxiety which motivates them to represent the risk in a specific manner. A deep-seated mental process used to relieve anxiety, is splitting an unconscious defence mechanism, associated with embracing the self-pleasing experiences and feelings, and projecting outward the self-displeasing feelings. A universal life force motivates this protective process and it comes to the surface whenever dangers are represented. Hence, rather than focusing on intrapersonal cognitive processes with their limitations, SRT emphasizes the precise and multifaceted content associated with general thought processes regarding risk and emphasizes media-mind link, proceeding from 'I' to 'we' (Joffe, 2003). In the context of climate change, since it is also a socially represented phenomenon, greatly influenced by media, individual level risk assessment has limited meaning. Hence, SRT can prove to be a useful proposition to address public risk perception overtaking logical positivism associated with scientific studies of climate change risk. However, cross-cultural empirical research is needed to substantiate the same.

CLIMATE RISK APPRAISAL AND PERCEPTION

How can we encourage all the disbelievers and people, who don't really feel threatened by climate change, to do (and pay for) the mitigation/coping efforts? Risk appraisal can be characterized as an individual's ability to recognize a risk and its possible harm to substances that are vital for human existence. It has two sub-components respectively, *perceived probability* (Anticipation of the human being susceptible to risk) and *perceived severity* (Assessing the effects of the danger associated with the crucial concerns of the person if the danger actually exists) (Grothmann & Patt, 2005).

Risk perception is found to be one of the motivating factors among other factors analysed against climate change adaptation behaviour and strongly associated with people's intentions to adapt (van Valkengoed & Steg, 2019). These are considered as the underlying rational short-cuts adopted by people to assess the threats and benefits of a climatic hazard by retrieving the pool of associated positive and negative emotions and feelings, consciously or unconsciously (Oatley & Johnson-Laird, 1996).

Climate change risk assessment includes systematic study of the implications, likelihoods, and solutions to climate change effects, as well as ways to mitigate them under societal constraints (Adger et al., 2018).

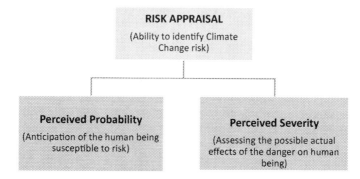

Fig. 2.2 Sub-components of risk appraisal

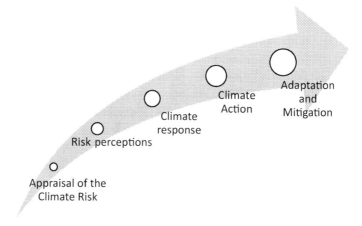

Fig. 2.3 Progression of climate risk appraisal and adaptation

It is the likelihood of human beings to perceive the possible risk associated with the changing climate. To respond against a threat, a person first recognizes a threat, known as risk appraisal, then makes a perception of it through cognitive mechanisms and affective concerns, and ultimately begins to act against reducing that particular threat which results in process of climate response, behavioural action at conative level and finally converging into adaptation and mitigation in context to climate change.

Cognitive mechanisms, especially the way information related to climate change and associated threat is processed in brain, in terms of probability as well as severity, plays a significant role in risk perception. In a study, effect of information on perceived risk to individuals was studied, with the aim to emphasize some factors supporting or hindering participation in adaptation behaviour. It was based on a psychological framework called the Protection Motivation Theory. It was observed, that higher perceived risk contributed to substantially higher incentive for adaptation. While comparing with the economic context, people who see climate change as a major stressor would end up investing more on climate conservation activities (Osberghaus et al., 2010). The manner in which people perceive the risk, plays a major role in shaping climate change policy and management response systems (Slovic, 2000). Environmental attitude help resolve the effect on ecological actions of both values (altruistic values and self-improvement ideals) and risks (perceived environmental challenges). Milfont et al. (2010) informed, that perceived personal threat of environmental problems is a strong predictor of ecological behaviour. The study found that both values and perceived perceptions of environmental risks could be used to foster ecological behaviour and resolve environmental concerns. In another study, beliefs about climate change, climate risk, attitudes and climate change perceptions of US farmers were evaluated against the option of effective adaptation approaches. Findings revealed that perceptions about climate risk of their farm and attitudes towards adaptation and innovation were critical determinants of adaptation process (Mase et al., 2017).

Risk perception operates at scientific level as well as intuitive and experiential level (Garvin, 2001; Kraus et al., 1992). Scientific level adopts the techniques of reasoning and valuation relying on probabilistic and mathematical description. Subjective evaluation of risks, which are intuitive and experiential in nature, relies less on measurement of associated injury and death. It rather evaluates risks in a more qualitative manner varying predictably by demographic and psychological attributes, individual personal and historical experiences, and social context (Jasanoff et al., 1998) as reflected in Fig. 2.4.

Risk perception across the social dimension forms comprehension of risk perceptions within society, views of various communities and conflicts, examines study and coordination goals and assesses risk management trade-offs (Davidson et al., 2003). People's views on risk or adaptation capacity in response to climate change are formed and informed by

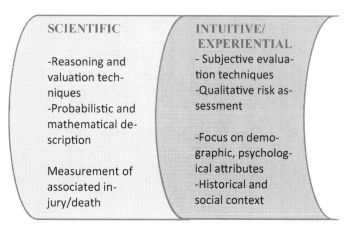

Fig. 2.4 Levels of operation of risk perception

what they read about climate change in the media, from peers, family, neighbours, or public authorities as revealed by based on 'social amplification of risk' theory that addresses the influence of the broader societal, institutional, and cultural context upon climate change perception and behaviour (Kasperson et al., 1992).

Determinants of Risk Perception

Public risk perceptions can fundamentally compel or constrain political, economic, and social action to address risks. Public support or opposition to climate policies (e.g., treaties, regulations, taxes, subsidies) are greatly influenced by public perceptions of the risks and dangers posed by global climate change. Perceptions about risk are determined by an array of factors which have been evaluated in various research studies. Americans contribute to a moderate level of perception of climate change threats, allowing them to actively advocate national and international climate change mitigation policies. Leiserowitz (2006) conducted a survey of the US public and found that Americans who have moderate climate change risk perceptions, strongly support a variety of national and international policies to mitigate climate change, and strongly oppose several carbon tax proposals. Drawing on the theoretical distinction between analytic and experiential decision-making, this study found that American risk perceptions and policy support are strongly influenced by experiential factors, including affect, imagery, and values, and demonstrates that public responses to climate change are influenced by both psychological and socio-cultural factors.

Another work, includes interrelated role of three factors, namely personal experience, affect and risk perception, investigated against cognitive mechanisms in defining climate change perceptions, revealing that personal experience with extreme weather is the best predictor of climate change perceptions that lead to risk perception as a predictor of affect; along with a reciprocal influence of risk perception and affect with each other (van der Linden, 2014). Further, a comprehensive social-psychological model of climate change risk perceptions was tested by incorporating cognitive, experiential and socio-cultural factors such as gender, political party, knowledge of the causes, impacts and climate change response, social norms, value orientations, affect and personal experience with extreme weather against climate change perceptions, as important predictors of climate change expectations (van der Linden, 2015).

Xie et al. (2019) extended the model of climate change risk perception by van der Linden (2015) to predict the risk perception of Australians and their willingness to engage in climate change mitigation behaviour. The major predictors were affect, mitigation response inefficacy, and descriptive norms with the prominence of affective, cognitive, and socio-cultural factors in explaining the variance of behavioural willingness.

From all the above factors, cognitive level understanding and knowledge related to the causes and impacts of the climate change was found to be an essential motivator to participate in protection strategies which is one of the significant factor described by Kroemker and Mosler (2002), that influence individual's protection capacity. Review of several dimensions of risk management and perception makes it evident the crucial role of these entities in climate change adaptation which can create new conduits for wisdom critical for research and developmental policies and initiatives to strengthen mitigation and adaptation processes (Mase et al., 2017; Negi et al., 2017).

Perceiving climate change as a risk is also a matter of what type of risk people perceive it as. Climate change is seen as a slow and gradual process that has taken place over years and years. Though the impacts are very much obvious in recent times, still the required urgency is not experienced. The perception of risk is highest at the micro level when a person is threatened by some specific impact of changing climate on him or his/her immediate relationship circle. It gradually reduces its strength as it extends its circle from person to town to state to country to globe.

Some of the perceptions in respect to climate change across the world as documented in various studies, depicted that local people who were well aware of changes in hydro-climatic parameters in that area, made adjustments to their aquaculture practices and these perceptions were influenced due to frequent adverse impacts of climate stressors on their livelihood activities (Shameem et al., 2015). In another case, people of the Kano state, Nigeria posed major threats to its agriculture systems; river and dam adjacent communities; the health sector; the transport and aircraft sectors; and urban ecosystems that are well-observed by them. Such risks are not appropriately handled by their institutions operating in the area of adaptation to climate risks, pointing to a poor potential for adaptation and weak governance (Aliyu & Amadu, 2017).

If people can foresee a potential fatality or damage, uncertainty regarding undesirable and uncontrollable consequences, they are considered as cognitively aware about climate change and associated risk. That is the reason why scientists and climatologists are much more worried about the future of planet due to climate change, than the public.

Risk Perception Processes

Social-behavioural science, has contributed extensively to risk perception research initiated by (Slovic et al., 1982), and on risk representation by (Joffe, 2003) to conceptualize and assess CC risk appraisal and adaptation. Evidence from cognitive psychology and social psychology indicates that risk perceptions, in a broad range of domains, are much more influenced by associative and affect-driven processes than by analytic processes (Chaiken & Trope, 1999; Epstein, 1994). Human information processing is done by either associative process or analytic process. Associative processing is quite simpler while analytic processing is quite complex. A common person can understand any phenomenon like climate change, through associative process while climatologists and scientists base their interpretations upon analytic processing. Analytic processing is a bit slow and needs conscious efforts, while associative processing is fast and automatic requiring less effort. If risk perceptions were driven mostly or exclusively by analytic considerations of consequences, they would not be influenced by the way a particular hazard is labelled (Fig. 2.5).

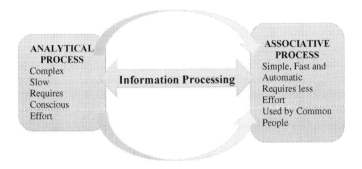

Fig. 2.5 Processes guiding climate change risk perception

For any person, climate change risk perception depends upon the outcome of either of the two information processing systems which generally operate in parallel and in close association. However, if the outcomes from both the systems are at odds and if one of the systems has to prevail, the associative process rules over the analytic system (Grothmann & Patt, 2005) just like the case of coastal population continuing to live near the coast despite being well aware of the risks posed by unforeseen and unpredictable coastal adversities. Despite knowing that this action puts their life in peril, they cannot avoid living in coastal vicinity.

The reason behind is that gradual climatic changes do not elicit the required anxiety which people have because of a feeling of lack of control over catastrophic consequences and hence climate change is not perceived as an 'immediate risk'. For much of the public in both east as well as west, climate change is still an "unknown risk" which is new and has unforeseeable consequences and therefore hard to believe or feel at risk about it. The greater the exposure to hazards, the more readily the adverse consequences are recognized by people. For example, the potential catastrophes from climate change, depicted in the film '*The Day after Tomorrow*' can generate and raise strong risk reactions and responses. Therefore, if climate change is interpreted as rapid, it is more likely to generate feelings of risk than if it is projected as too distant to adversely affect the society at any part of the world.

Now the issue is 'whether it is appropriate to create a fear psychosis about climate change and use it as a strategy to perceive risk to promote micro level pro-climate behaviour'? Genuine data-based public information must be provided to public which may create a state of arousal to perceive the climate change risk and behave responsibly. In this regard, drive reduction model of (Higbee, 1969) is worth to be cited. He states that fear appeal is a potent strategy to change the attitudes. It is also being used by media to promote awareness and concern about certain issues like alcoholism, smoking or so which public otherwise ignores. The same model can be adapted to modify climate change perceptions too by generating fear appeal as exhibited in various movies on this theme like '*The Day after Tomorrow*' mentioned previously. The movie generates a kind of fear in the mind of viewers that this may happen to our country too, which increases their level of arousal about climate change and leaves them thinking that adopting some action strategies is actually important. However, one cannot be left in a stage of anxiety. So reassurance is needed to stabilize the aroused system which may consequently lead to modified climate change risk perceptions with possible adoption of action strategies. Maddux and Rogers (1983) explored the effect of fear appeals on persuasion by testing the combination of protection motivation theory and self-efficacy theory. It was found that intention to adopt the recommended behaviour and efficacy of coping response increases with the probability of a threat's occurrence, supporting for self-efficacy expectancy as the component of protection motivation theory.

Subsequently, O'Neill and Nicholson-Cole (2009), using visual and iconic representations of climate change for engagement of people, demonstrated that in spite of the potential of these representations for enticing the responsiveness to climate change temporarily, fear is generally not much effective to motivate long-term personal engagement. Non-threatening visual imagery and icons that link to human emotions and concerns of daily life in the context of this macro-climatic scenario are likely to be more engaging. Therefore, mere fear is not enough to procreate and maintain climate friendly behaviour. Many a times, when individual freedom to act is curtailed, it becomes counterproductive to desired action as people become simply complacent considering it as one of the media gimmicks to catch hold of publicity. This aspect is further elaborated at length in the subsequent chapters on climate change communication and behavioural transformation, while explaining the role of emotions.

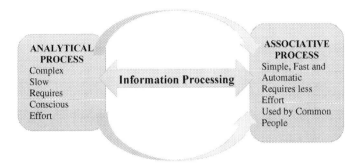

Fig. 2.5 Processes guiding climate change risk perception

For any person, climate change risk perception depends upon the outcome of either of the two information processing systems which generally operate in parallel and in close association. However, if the outcomes from both the systems are at odds and if one of the systems has to prevail, the associative process rules over the analytic system (Grothmann & Patt, 2005) just like the case of coastal population continuing to live near the coast despite being well aware of the risks posed by unforeseen and unpredictable coastal adversities. Despite knowing that this action puts their life in peril, they cannot avoid living in coastal vicinity.

The reason behind is that gradual climatic changes do not elicit the required anxiety which people have because of a feeling of lack of control over catastrophic consequences and hence climate change is not perceived as an 'immediate risk'. For much of the public in both east as well as west, climate change is still an "unknown risk" which is new and has unforeseeable consequences and therefore hard to believe or feel at risk about it. The greater the exposure to hazards, the more readily the adverse consequences are recognized by people. For example, the potential catastrophes from climate change, depicted in the film '*The Day after Tomorrow*' can generate and raise strong risk reactions and responses. Therefore, if climate change is interpreted as rapid, it is more likely to generate feelings of risk than if it is projected as too distant to adversely affect the society at any part of the world.

Now the issue is 'whether it is appropriate to create a fear psychosis about climate change and use it as a strategy to perceive risk to promote micro level pro-climate behaviour'? Genuine data-based public information must be provided to public which may create a state of arousal to perceive the climate change risk and behave responsibly. In this regard, drive reduction model of (Higbee, 1969) is worth to be cited. He states that fear appeal is a potent strategy to change the attitudes. It is also being used by media to promote awareness and concern about certain issues like alcoholism, smoking or so which public otherwise ignores. The same model can be adapted to modify climate change perceptions too by generating fear appeal as exhibited in various movies on this theme like '*The Day after Tomorrow*' mentioned previously. The movie generates a kind of fear in the mind of viewers that this may happen to our country too, which increases their level of arousal about climate change and leaves them thinking that adopting some action strategies is actually important. However, one cannot be left in a stage of anxiety. So reassurance is needed to stabilize the aroused system which may consequently lead to modified climate change risk perceptions with possible adoption of action strategies. Maddux and Rogers (1983) explored the effect of fear appeals on persuasion by testing the combination of protection motivation theory and self-efficacy theory. It was found that intention to adopt the recommended behaviour and efficacy of coping response increases with the probability of a threat's occurrence, supporting for self-efficacy expectancy as the component of protection motivation theory.

Subsequently, O'Neill and Nicholson-Cole (2009), using visual and iconic representations of climate change for engagement of people, demonstrated that in spite of the potential of these representations for enticing the responsiveness to climate change temporarily, fear is generally not much effective to motivate long-term personal engagement. Non-threatening visual imagery and icons that link to human emotions and concerns of daily life in the context of this macro-climatic scenario are likely to be more engaging. Therefore, mere fear is not enough to procreate and maintain climate friendly behaviour. Many a times, when individual freedom to act is curtailed, it becomes counterproductive to desired action as people become simply complacent considering it as one of the media gimmicks to catch hold of publicity. This aspect is further elaborated at length in the subsequent chapters on climate change communication and behavioural transformation, while explaining the role of emotions.

Hence, there is a need to take into consideration the barriers to risk perception and adaptation efforts and work for intrinsically motivating strategies like non-threatening visual and auditory representations to promote human engagement with climatic issues in a sustained manner and resulting climate consciousness.

BARRIERS TO CLIMATE RISK PERCEPTION AND SUSTAINABILITY

There are several factors in risk perception domain which are still present in society acting as the obstacles towards progression in climate change resilience process. One such barrier is the belief that scientific evidence and information are not so precise due to technical snags in complex geo-climatic areas. Hence, risk perceptions offer a stronger solution and tend to improve such scientific expertise. A need for robust institutional support is recommended besides establishing collaboration between organizations at regional, state, and national level. It is currently lacking in various parts of the eastern world particularly in India and Nepal (Uprety et al., 2017). Another constraint considers risk perceptions as a diverse set of system constituting of psychological, sociological, and cultural dimensions of risk-associated behaviour (van der Linden, 2015, 2017). However, there is limited empirical research pertaining to the psychological context of risk perception (Grothmann & Patt, 2005), which is pertinent to overcome various constraints associated with climate risk appraisals.

In a similar context, conserving energy is crucial to responsible consumption and resultant pro-climate action. It is considered as the individual reflection of sustainability efforts. However, there is a common tendency, especially in youth, to be reluctant in adopting energy efficient behaviours though pro-ecological attitudes are present. Dursun et al. (2019) explored the dynamics of such behaviour and attributed it to various psychological barriers, especially denial mechanisms and shrinking of moral obligations. Besides psychological denial, there are a range of other interfering factors/barriers which come in between climate change belief and climate action like social norms, limited information, limited control over events, long-held habits, limited trust over scientific community and limited resources. Another barrier may be perceived psychological distance. People mostly perceive that climate change affects distant countries, not my country, other species, not humans, people in future, not my generation, other segments of society, not my society and similar other distant variables. It substantially discourages them to perceive associated risk and resultant pro-climate action.

Development of independent environmental cognitions through education can break down the psychological barriers and enable the climate resilience and subsequent adaptation action as explained further.

CLIMATE RISK, RESILIENCE, AND ADAPTATION

Adaptation is a three-level framework that meddles in development, with the first being adaptation through enabling resilience, the second comprising transition, and the third being transformation (Pelling, 2010). Discussing about the resilience term in the first level of adaptation deals with the system's degree of elasticity, or its capacity to rebound or bounce back after being subjected to climate stress or shock and comprises of degree of flexibility and stability of specific functions as being its indicators. In addition, IPCC defines resilience as 'The ability of a social or ecological system to absorb disturbances while retaining the same basic structure and ways of functioning, the capacity for self-organization, and the capacity to adapt to stress and change'.[5]

Resilient people foresee risks, reduce vulnerability to those risks, respond effectively to threats, and recover faster, allowing them to respond to the next threat more quickly (Ebi, 2011). Further, Individual resilience is dependent on the ability to identify localized climate risk (Climate risk appraisal), the motivation to learn and plan for these prospective climate risks, and general appraisals of personal adaptive capacities (Smith et al., 2012).

Adaptation to build resilience simply seeks modifications that will allow existing functions and practices to continue while neglecting fundamental assumptions and structural inequality in society. Transition acts as a conduit between resilience and transformation, encompassing governing regimes through activities that attempt to express complete rights rather than systemic changes. Reforms in underlying political-economic regimes and accompanying cultural ideologies on progress, security, and risk reflect the deepest level of adaptation, which is signified by transformation (Pelling, 2010). The transformation level will comprise of Adaptation

[5] IPCC (2008) Glossary of Terms for Working Group II. Accessed 10 May 2010 from http://www.ipcc.ch/pdf/glossary/ar4-wg2.pdf.

behaviour which includes changes in habits, trend adjustments, and enhanced resilience and is more proactive and has a longer time horizon (Singh et al., 2016). A lack of risk appraisal and adaptation appraisal results into a cognitive barrier to adaptive action (Grothmann & Patt, 2005).

In a study, the role of community resilience and risk appraisal on climate change adaptation behaviour was highlighted. Residents of a small township (which was built on flood prone area and exposed to flooding risk and water scarcity) in Chennai, a southern coastal metropolitan city of India, reflected positive influence of Five core dimensions of community resilience (place attachment, trust, social networks, collective efficacy, and social support) on adaptation actions. It was found that social networks along with community's self-efficacy can increase risk awareness which can finally result into enhanced number of adaption measures (Haas et al., 2021).

Conclusion

The climate change perception is represented as a joint function of the objective environmental conditions (for example, population density, temperature, pollution levels) and the individual characteristics of the person (for example, adaptation level, previous experience with climatic events etc.). If the perceived environment is outside the individual's optimal range (for example, if it is over stimulating, contains too many stressors, constrains behaviour, or offers insufficient resources), stress is experienced which, in turn, elicits coping. If the attempted coping strategies to climate change are successful, adaptation and/or habituation occurs, however, it is possibly followed by aftereffects such as fatigue and reduced ability to cope with the next climate stressor. Positive cumulative aftereffects would include a degree of learning about how to cope with the next occurrence of undesirable environmental stimulation/climatic event. Systematic research regarding climate risk appraisal and perception for future extreme weather events and its Positive communication will be needed in diverse geo-climatic regions/countries, to take it further.

If adequate and informative climate-centric media reports reach people that aim at providing knowledge about climate change consequences at the ground level, besides providing the global perspective, there is a greater probability of cognitive assimilation and affective internalization by the target population. This is because people form judgements and emotions about any happening such as climate change in the context of their values and their judgements and emotions do depend upon whether they give priority to moral or to utilitarian outcomes in the given context. It is simply impossible to expect the public to be aroused or concerned about climate change or anticipate effective action on their part by simply providing them with climate change information as being done by IPCC since decades. Climate change concern can be translated into action only when people realize their personal accountabilities for the issue. This will not come until and unless people are made aware of what specific contributions they can make and the initiatives they can take on personal level to combat climate change impacts. For generating individual action, opportunities and incentives are required. These can be very effectively created with the help of communities and organizations working for the cause of climate change. The dedicated chapter on dynamics of behavioural transformation explains it further. To support the individuals in their attempts at preserving the environment and more specifically in their efforts to counter climate change, the organizations, communities, religious groups, and social organizations can engage in broader social efforts towards climate change. This would help individuals have the support and encouragement of neighbours and friends so that they do not feel alone in their actions to ward off climate change.

References

Adger, W. N., Brown, I., & Surminski, S. (2018). Advances in risk assessment for climate change adaptation policy. *Philosophical Transactions of the Royal Society A: Mathematical, Physical and Engineering Sciences, 376*(2121), 20180106.

Aliyu, A. A., & Amadu, L. (2017). Urbanization, cities, and health: The challenges to Nigeria—A review. *Annals of African Medicine, 16*(4), 149.

Biermann, F. (2002). *Global environmental change and the nation state: The scope of the challenge* (PIK Report), 1.

Chaiken, S., & Trope, Y. (1999). *Dual-process theories in social psychology*. Guilford Press.

Davidson, K., Daly, T., Arber, S., & Ginn, J. (2003). Exploring the social worlds of older men. In S. Arber, K. Davidson, & J. Ginn (Eds.), *Gender and ageing: Changing roles and relationships* (pp. 168–185). Open University Press.

Dursun, İ., Tümer Kabadayı, E., & Tuğer, A. T. (2019). Overcoming the psychological barriers to energy conservation behaviour: The influence of objective and subjective environmental knowledge. *International Journal of Consumer Studies, 43*(4), 402–416.

Duveen, G. (2000). The power of ideas. In S. Moscovici (Ed.), *Social representations: Explorations in social psychology* (pp. 1–17). Polity Press.

Ebi, K. L. (2011). Resilience to the health risks of extreme weather events in a changing climate in the United States. *International Journal of Environmental Research and Public Health, 8*(12), 4582–4595.

Ehrhardt-Martinez, K. (2011). Changing habits, lifestyles and choices: The behaviours that drive feedback-induced energy savings. In *Proceedings of the 2011 ECEEE Summer Study on Energy Efficiency in Buildings, Toulon, France, 2011* (pp. 6–11).

Epstein, S. (1994). Integration of the cognitive and the psychodynamic unconscious. *American Psychologist, 49*(8), 709.

Garvin, T. (2001). Analytical paradigms: The epistemological distances between scientists, policy makers, and the public. *Risk Analysis, 21*(3), 443–456.

Grothmann, T., & Patt, A. (2005). Adaptive capacity and human cognition: The process of individual adaptation to climate change. *Global Environmental Change, 15*(3), 199–213.

Haas, S., Gianoli, A., & Van Eerd, M. (2021). The roles of community resilience and risk appraisal in climate change adaptation: The experience of the Kannagi Nagar resettlement in Chennai. *Environment and Urbanization*. https://doi.org/10.1177/0956247821993391

Hackmann, H., Moser, S. C., & Clair, A. L. S. (2014). The social heart of global environmental change. *Nature Climate Change, 4*(8), 653–655.

Higbee, K. L. (1969). Fifteen years of fear arousal: Research on threat appeals: 1953–1968. *Psychological Bulletin, 72*(6), 426.

Iturriza, M., Labaka, L., Ormazabal, M., & Borges, M. (2020). Awareness-development in the context of climate change resilience. *Urban Climate, 32*, 100613.

Jasanoff, S., Wynne, B., Buttel, F., Charvolin, F., Edwards, P., Elzinga, A., Haas, P., Kwa, C., Lambright, W., & Lynch, M. (1998). Science and decision-making. In S. Rayner & E. Malone (Eds.), *Human choice and climate change, vol 1: The societal framework* (pp. 1–87). Battelle Press.

Joffe, H. (2003). Risk: From perception to social representation. *British Journal of Social Psychology, 42*(1), 55–73. https://doi.org/10.1348/014466603763276126

Joffe, M. M., & Rosenbaum, P. R. (1999). Invited commentary: Propensity scores. *American Journal of Epidemiology, 150*(4), 327–333.

Kasperson, R. E., Golding, D., & Tuler, S. (1992). Social distrust as a factor in siting hazardous facilities and communicating risks. *Journal of Social Issues, 48*(4), 161–187.

Kraus, N., Malmfors, T., & Slovic, P. (1992). Intuitive toxicology: Expert and lay judgments of chemical risks. *Risk Analysis, 12*(2), 215–232.

Kroemker, D., & Mosler, H.-J. (2002). Human vulnerability—Factors influencing the implementation of prevention and protection measures: An agent based approach. In K. W. Steininger & H. Weck-Hannemann (Eds.), *Global environmental change in alpine regions: Impact, recognition, adaptation, and mitigation* (pp. 95–114). Cheltenham: Edward Elgar.

Leiserowitz, A. (2006). Climate change risk perception and policy preferences: The role of affect, imagery, and values. *Climatic Change, 77*(1–2), 45–72. https://doi.org/10.1007/s10584-006-9059-9

Maddux, J. E., & Rogers, R. W. (1983). Protection motivation and self-efficacy: A revised theory of fear appeals and attitude change. *Journal of Experimental Social Psychology, 19*(5), 469–479.

Mase, A. S., Gramig, B. M., & Prokopy, L. S. (2017). Climate change beliefs, risk perceptions, and adaptation behavior among Midwestern US crop farmers. *Climate Risk Management, 15*, 8–17.

Milfont, T. L., Sibley, C. G., & Duckitt, J. (2010). Testing the moderating role of the components of norm activation on the relationship between values and environmental behavior. *Journal of Cross-Cultural Psychology, 41*(1), 124–131. https://doi.org/10.1177/0022022109350506

Moscovici, S., & Lage, E. (1976). Studies in social influence III: Majority versus minority influence in a group. *European Journal of Social Psychology, 6*(2), 149–174.

Negi, V. S., Maikhuri, R. K., Pharswan, D., Thakur, S., & Dhyani, P. P. (2017). Climate change impact in the Western Himalaya: People's perception and adaptive strategies. *Journal of Mountain Science, 14*(2), 403–416.

Oatley, K., & Johnson-Laird, P. N. (1996). *The communicative theory of emotions: Empirical tests, mental models, and implications for social interaction.*

O'Neill, S., & Nicholson-Cole, S. (2009). "Fear won't do it": Promoting positive engagement with climate change through visual and iconic representations. *Science Communication, 30*(3), 355–379. https://doi.org/10.1177/1075547008329201

Osberghaus, D., Dannenberg, A., Mennel, T., & Sturm, B. (2010). The role of the government in adaptation to climate change. *Environment and Planning C: Government and Policy, 28*(5), 834–850.

Pelling, M. (2010). *Adaptation to climate change: From resilience to transformation.* Routledge.

Rishi, P., & Mudaliar, R. (2014). Climate stress, behavioural adaptation and subjective well being in coastal cities of India. *American Journal of Applied Psychology, 2*(1), 13–21.

Rogers, R. W. (1975). A protection motivation theory of fear appeals and attitude change1. *The Journal of Psychology, 91*(1), 93–114.

Rogers, R. W. (1983). Cognitive and psychological processes in fear appeals and attitude change: A revised theory of protection motivation. In J.

T. Cacioppo & R. E. Petty (Eds.), *Social psychophysiology: A sourcebook* (pp. 153–176). Guilford Press.

Shameem, M. I. M., Momtaz, S., & Kiem, A. S. (2015). Local perceptions of and adaptation to climate variability and change: The case of shrimp farming communities in the coastal region of Bangladesh. *Climatic Change, 133*(2), 253–266.

Shiota, M. N., Papies, E. K., Preston, S. D., & Sauter, D. A. (2021). Positive affect and behavior change. *Current Opinion in Behavioral Sciences, 39,* 222–228. https://doi.org/10.1016/j.cobeha.2021.04.022

Singh, C., Dorward, P., & Osbahr, H. (2016). Developing a holistic approach to the analysis of farmer decision-making: Implications for adaptation policy and practice in developing countries. *Land Use Policy, 59,* 329–343.

Slovic, P. E. (2000). *The perception of risk.* Earthscan Publications.

Slovic, P., Fischhoff, B., & Lichtenstein, S. (1982). Why study risk perception? *Risk Analysis, 2*(2), 83–93. https://doi.org/10.1111/j.1539-6924.1982.tb0 1369.x

Smith, J. W., Anderson, D. H., & Moore, R. L. (2012). Social capital, place meanings, and perceived resilience to climate change. *Rural Sociology, 77*(3), 380–407.

Stott, P. A., & Sullivan, S. (2000). *Political ecology: Science, myth and power.* Arnold.

Takala, M. (1991). Environmental awareness and human activity. *International Journal of Psychology, 26*(5), 585–597. https://doi.org/10.1080/002075991 08247146

Uprety, Y., Shrestha, U. B., Rokaya, M. B., Shrestha, S., Chaudhary, R. P., Thakali, A., Cockfield, G., & Asselin, H. (2017). Perceptions of climate change by highland communities in the Nepal Himalaya. *Climate and Development, 9*(7), 649–661.

van der Linden, S. (2014). On the relationship between personal experience, affect and risk perception: The case of climate change. *European Journal of Social Psychology, 44*(5), 430–440.

van der Linden, S. (2015). The social-psychological determinants of climate change risk perceptions: Towards a comprehensive model. *Journal of Environmental Psychology, 41,* 112–124. https://doi.org/10.1016/j.jenvp.2014. 11.012

van der Linden, S. (2017). Determinants and measurement of climate change risk perception, worry, and concern. In S. van der Linden (Ed.), *Oxford research encyclopedia of climate science.* Oxford University Press. https://doi.org/10. 1093/acrefore/9780190228620.013.318

van Valkengoed, A. M., & Steg, L. (2019). Meta-analyses of factors motivating climate change adaptation behaviour. *Nature Climate Change, 9*(2), 158–163. https://doi.org/10.1038/s41558-018-0371-y

Weaver, C. P., Mooney, S., Allen, D., Beller-Simms, N., Fish, T., Grambsch, A. E., Hohenstein, W., Jacobs, K., Kenney, M. A., Lane, M. A., Langner, L., Larson, E., McGinnis, D. L., Moss, R. H., Nichols, L. G., Nierenberg, C., Seyller, E. A., Stern, P. C., & Winthrop, R. (2014). From global change science to action with social sciences. *Nature Climate Change*, 4(8), 656–659. https://doi.org/10.1038/nclimate2319

Xie, B., Brewer, M. B., Hayes, B. K., McDonald, R. I., & Newell, B. R. (2019). Predicting climate change risk perception and willingness to act. *Journal of Environmental Psychology*, 65, 101331.

CHAPTER 3

Climate Change and Sustainability Communication—A Socio-Psychological Analysis

INTRODUCTION

Global issues of Sustainability are marked with a high degree of convolution and ambiguity making it challenging for common people to comprehend and act upon as desired for achieving the Sustainable Development Goals (SDGs). 'SDGs were announced in 2015, to address the challenges of the economic divide, information asymmetry, dearth of accessible services for common people, digital divide, and climate change'.[1] One of the leading challenges for successful achievement of SDGs is 'how to communicate the SDGs and how to build the stake of people who can be the change-makers'.[2] Hence, the role of communication is especially critical to cope with these issues (Newig et al., 2013) and it becomes further crucial during this challenging scenario of COVID-19. Luhmann (1997) has also aptly remarked 'Society is unthinkable without communication, but communication is also unthinkable without society.

---

[1] https://www.local2030.org/library/674/SDG-Vision-for-Digital-India.pdf.

[2] https://www.comminit.com/content/communicating-sustainable-development-goals-toolkit-community-radio.

© The Author(s), under exclusive license to Springer Nature Singapore Pte Ltd. 2022
P. Rishi, *Managing Climate Change and Sustainability through Behavioural Transformation*, Sustainable Development Goals Series, https://doi.org/10.1007/978-981-16-8519-4_3

Communication is the basic operation that produces and reproduces societies'.

Human interface with the natural and social environment in a responsible manner is what climate change and sustainability communication calls for. It is a process of social indulgence dealing with the possible causes of climate change and unsustainable behaviour in diverse fields and looking for their possible solutions, which are acceptable to people. Actively involving people in the process of steering Sustainable Development through development of affective/emotional concerns is equally pertinent.

For such active involvement, what is necessitated is wider access to information, as what is globally available on world-wide-web is not necessarily accessible to local people. Narendra Modi, Prime Minister of India remarks—'The sustainable development of one-sixth of humanity will be of great consequence to the world and our beautiful planet'.[3] With the COVID-19 pandemic and associated loss to individual incomes as well as global economies, the disparities and inequalities in information access are further widened in developing countries, and the 'digital divide', expands the gap further between the 'front-runners' and 'back-seaters' of climate change and sustainability communication. This trend of widening division in different societies reinforces the already strained economies causing distortions in socio-cultural and ecological milieu of almost all the regions. It warrants assuming responsibility on the part of individuals, institutions as well as governance systems worldwide to redesign their relationships with each other and the natural environment.

A psychological and social process of understanding among diverse population groups with a vision for sustainability and a solution-oriented mindset can possibly make it happen. The chapter analyses and evaluates climate change and sustainability communication from a transdisciplinary perspective integrating communication models and concepts from the disciplines of philosophy, psychology, sociology, and political science. Further, it draws examples from diverse fields of understanding and proposes its application as a management tool.

---

[3] https://in.one.un.org/page/sustainable-development-goals/.

Framework of Sustainability Communication

Mack Bhatia remarks 'Sustainability communication is a strategic approach to engage stakeholders who either are looking up to you as consumers, or institutional investors with sustainability portfolio, willing to invest in you. It is extremely important for all stakeholders, but specifically for investors to know that you are truly on a sustainable journey'.[4] Climate change and sustainability communication is not just limited to communicating information and generating awareness about sustainability concerns at the cognitive level of understanding. It is to analytically appraise and bring together a thoughtful perspective of the human–environment interrelationship into *social discourse* (Godemann & Michelsen, 2011). Authors have also presented some of the distinctive features of Sustainable communication which, in an adapted form, are outlined in the figure below (Fig. 3.1).

In simple terms, reflective thinking is a process of taking a pause, consciously thinking about the issue in question, connecting it to already existing insights and experiences, questioning and critically examining your own thoughts, values, assumptions and stereotypes and finally drawing practical applications from it through examining our composite roles in relation to significant others, following a pluralistic and inclusive approach.[5] In other words, reflective thinking is marked with conscious thinking, perennial learning for self-improvement with insights and experiences and self-empowerment to make necessary behavioural changes/corrective actions in regard to the issue of concern. Sustainable Communication starts with critical and reflective thinking framework exploring the issues of sustainability in business or societal context; why they deserve public attention and need to be communicated; and how to communicate them (including the content and process of communication) in a manner so that the issue pertaining to sustainability gets public attention and is duly addressed. Inclusive approach to sustainability communication is highly desirable to have wider acceptance among diverse population groups in order to create intrinsic social value. Social

---

[4] What is Sustainability Communication? Is it different than ESG Communication (the sustainability.io).

[5] https://learningforsustainability.net/reflective-practice/.

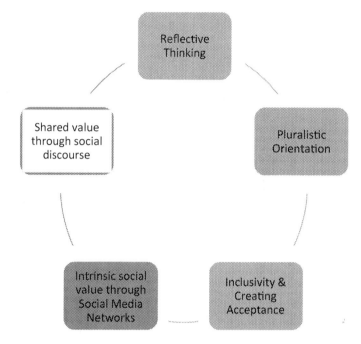

Fig. 3.1 Distinctive features of sustainability communication

media networks play a significant role in this regard for creation of shared value through social discourse.

'Critical thinking and sustainable development is an approach to meaningful dialogue for social, economic, political and environmental problem-solving and decision-making……..the practice of engaging in analytical dialogue and problem-solving mechanisms, through active mental and emotional inquiry, for the transformation of individuals, communities and institutions' (Minott et al., 2019). The framework of sustainability communication can be broadly classified into three subcategories: corporate sustainability communication; sustainable consumption communication (Bilharz & Schmitt, 2011); and climate change communication (Newig, 2011). All these subsystems are guided by human psyche and value systems that exist within the broader framework of socio-cultural subsystems. Following model represents this framework in a systematic manner (Fig. 3.2).

Fig. 3.2 Sustainability communication framework

Corporate Sustainability Communication

The term sustainability communication is universally used to reflect the corporate sustainability policy with the intention to make it evident to the stakeholders, the dedicated commitment of organization regarding pursuance of sustainable development goals (SDGs) and the steps/process they are following for sustainable transition.[6] Leaders play an important role in fostering a sustainability-focused culture to achieve positive employee-level and organizational-level sustainability performance (Galpin et al., 2015). Most of the time, the role of sustainability leadership blends well with corporate social responsibility (CSR) initiatives of the organization. With the increasing global emphasis on SDGs, having a sustainability policy and communicating it to the competitive world, is imperative for corporate organizations. It was possible only due to contentious discussions and deliberations on the issues of ecologically unsustainable development and practices in the mass media in order to generate pressure on institutions and organizations to take responsible action (Brand, 2006).

Sustainability strategy of every corporate organization is closely aligned to their communication practices. Limited and insufficient internal communication makes implementation of changes difficult which are required for making a sustainable organization (Genç et al., 2017).

---

[6] https://www.thesustainability.io/what-is-sustainability-communication-esg.

Fig. 3.3 Sustainability communication-internal and external

Similarly, external communication with stakeholders regarding sustainability strategies is equally important looking at the global trends towards sustainability. Therefore, integration of sustainability issues in business propositions has become a survival strategy for the corporate world.

As reflected in the figure below, it is imperative that companies demonstrate how much investments they are making in sustainable business strategies through communication strategies suiting the requirements of diverse stakeholders (Fig. 3.3).

A Shared Value Creation
Sustainability is a multidimensional concept closely aligned to societal communication. Hence, societal discourse is essential to legitimize sustainable development (Newig et al., 2013). In this context, shared values creation through participation of strategic stakeholders helps to assimilate multiple stakeholders. 'Creating shared value in the context of sustainability is a way to integrate stakeholders in business management'.[7] It is only possible through adoption of appropriate communication strategy for sustainable activities (López & Monfort, 2017). The notion of '*creating shared value*', was first given by Michel Porter and Mark Kramer emphasizing that economic value generation closely intersects with social issues and its enhancement is possible through identifying

---
[7] http://dx.doi.org/10.5772/intechopen.70177.

and addressing social problems.[8] They further elaborate that it is possible through the intersection of business needs, social needs/concerns, and business assets/expertise into a shared value for stakeholders. Engaging with key stakeholders through continuous dialogue to learn and establish a common platform can help create a strategic value of sustainability. In this manner, 'their sustainability agenda is better positioned to anticipate and react to economic, social, environmental, and regulatory changes as they arise'.[9]

López and Monfort (2017) conducted a study of seven reputed and recognized companies in their respective sectors, i.e., Philips (Netherlands), LG Electronics (Republic of Korea), Nestle (Switzerland), BMW group (Germany), IIPE (United States), Unilever (Netherlands), and Roche Pharmaceuticals (Switzerland). All the selected companies were having their sustainability strategy integrating Environment, Social & Governance (ESG) approach. Authors concluded that shared value creation can explain how 'value proposition' of the companies is integrated into their business strategy giving due regard and importance to their diverse groups of stakeholders for creation of an all-encompassing dialogue based approach. As a result, the companies succeeded in impacting the society through their CSR strategy following stakeholders' engagement approach to transform the communities and society in a measurable and impactful manner. It was possible through extensive and intelligent use of social media communication.

The concept of shared values is relevant not only from the corporate perspective but also from the community perspective, especially concerning achievement of SDGs. In order to elucidate and simplify SDGs and their 169 targets by 2030, a toolkit for community radio (CR) stations was designed by an Indian organization '*Seeking Modern Applications for Real Transformation* (SMART)' in collaboration with United Nations Educational, Scientific and Cultural Organization (UNESCO) and the United Nations Children's Fund (UNICEF). It was primarily meant for reporters, volunteers and community media practitioners in India containing evidence, endorsements, and strategies for undertaking community engagement activities to encourage local action in relation to

---

[8] https://richtopia.com/effective-leadership/michael-porter-shared-value/.

[9] https://hbr.org/2016/10/the-comprehensive-business-case-for-sustainability.

the SDGs. It has dedicated content for communicating SDG 11 (Sustainable cities and communities), SDG 12 (Sustainable consumption and production), and SDG 13 (Climate action), which is context relevant to engage with diverse population groups of India, taking into consideration the language barriers and having a wider reach through community radio stations.[10]

Sustainability as a Tool for Competitive Advantage
Today's business professionals are coming across multifaceted and unparalleled infusion of societal, ecological, and organizational developments which make sustainability-based management imperative for them. However, they still have a mistaken belief that it is going to be a costly affair for them which is not actually true. If we draw conclusions from business experience as well as academic research, sustainability, embedded in business practices, has the potential to leave a positive bearing on long-term business performance. From the perspective of the business world, sustainability is explained as the active initiatives/operations of the company which causes minimum harm to people and planet while being able to create the best value for stakeholders. While doing so, they should be able to focus on Environment, Social, and Governance (ESG) performance and create an environmental and social impact.[11] Broadly speaking, under the aegis of the term sustainability, a desirable connection with community and the society, an expansion of potential customers, and a long-term reduction in operational expenditures is embedded. Hence, through sustainability communication, companies can draw long-term competitive advantage (Doorley & Garcia, 2015; von Kutzschenbach & Brønn, 2006) along with enhancing the company's goodwill. There are empirical studies across countries to substantiate this claim. Hespenheide et al. (2010) reported that a substantial percentage of American consumers are driven by sustainability in general while half of them consider at least a single feature of sustainability while selecting a brand/company of their choice to buy products.

Gerard Escaler, an international marketing professional, writes in Forbes communication council five strategies which organizations can

---

[10] https://www.comminit.com/content/communicating-sustainable-development-goals-toolkit-community-radio.

[11] https://hbr.org/2016/10/the-comprehensive-business-case-for-sustainability.

adopt to stand apart from their contending counterparts[12]: Firstly, alignment and regular appraisal of SDGs with operational strategies of business to define which SDGs are most closely aligned to business and revenue model. Accordingly, they can look for opportunities to reinforce business models with embedded social and environmental impact. Secondly, focus on being innovative in the ways it is implemented, much beyond merely a marketing strategy. Thirdly, networking and partnerships with like-minded organizations/companies potentiate the impact. Fourthly, without close monitoring and measurement of sustainability initiatives, directions are unclear. Hence, through annual SDG reporting, sustainability initiatives can be quantified. Finally, scalability of sustainability initiatives for communities to move beyond a revenue generation/profit motive to magnify the impact which is possible through employee engagement and volunteering spread across key-impact areas.

Expansion of the sustainability programme into core areas of business operation can provide an edge over others in terms of market penetration, client and employee branding, and community development, giving distinct business identity along with giving a boost to a company's competitive advantage.

Sustainability as a Management Tool—Communication vs. Commitment

Baldassarre and Campo (2016) proposed a matrix to reflect the level of business sustainability in terms of commitment to a company's sustainable initiatives and the communication of such responsible and environmentally friendly initiatives to stakeholders. The matrix combines commitment 'being sustainable' vs communication 'appearing to be sustainable' into four diverse but dynamic conditions of transparency depending on the intention of sustainability as reflected in the figure below (Fig. 3.4).

Translucent companies are companies which tend to meet their sustainability goals but are not able to translate their achievements into good marketing campaigns. While the companies invest and embed sustainability into their system, the management fails to realize the importance of communicating the same to the stakeholders. As a result, they are unable to extract strategic advantage from their sustainability initiatives and there remains a gap between sustainability performance of the company

---

[12] https://www.forbes.com/sites/forbescommunicationscouncil/2020/09/09/transforming-sustainability-into-a-competitive-advantage/?sh=5c082c4f282e.

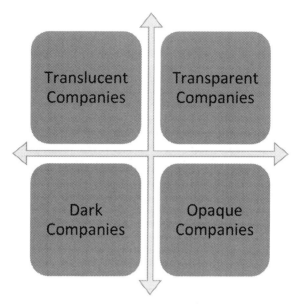

Fig. 3.4 Commitment vs. communication matrix[13]

and stakeholders' perceptions concerning their sustainability efforts. For example, tribal corporations in India have been sustainably harvesting minor forest products for decades and supplying them to FMCG giants in the country. But they have failed to communicate and quantify their approaches in the process of sustainable harvesting. Similarly, an Indian indigenous organization- TRIFED,[14] running development enterprises affecting the lives and livelihoods of tribal artisans and forest produce gatherers, has expanded sustainably over the last few years in terms of ever-increasing investments leading to '*sustainable value chain*' across the nation. It is very crucial to the booming development sector of India to have '*equitable, inclusive and sustainable growth*' but it is equally important to communicate the same to its stakeholders, to develop widespread sustainability perceptions and experience the benefits of their sustainable positioning.

[13] Adapted and produced with permission from Baldassarre and Campo (2016).
[14] https://trifed.tribal.gov.in/general-information-activities.

Transparent companies take advantage of their sustainability initiatives to gain a competitive advantage against their peers. All the sustainable activities are well marketed and communicated to engage the target customers. Definite metrics are set for each activity and are communicated well in the sustainability reports of the organization. IKEA derives its competitive advantage and sets industry standards via its sustainable practices. Targeting the millennial generation, IKEA has been successful in creating a brand image of a sustainable company. They have partnered with WWF, UNICEF, and other social organizations to cope with the environmental and social aspects of their business. IKEA has embedded sustainability across its value chain and procures majority of its wood certified from FSC (Forest Stewardship Council).[15] They have been constantly reporting on sustainability aspects and have developed an effective sustainability communication model. Adidas, one of the leading designers and manufacturers of shoes, clothes, and accessories has come up with the '*Run for the Oceans*' initiative. For every 1 kilometre run, the company claims to clean up 10 plastic bottles from the oceans. Adidas has also come up with products made up of plastics dumped into the oceans.[16] Thus, they are blending their sustainability initiatives with good marketing.

Dark companies are not aware of sustainability and its possible strategic advantage to their company. They also fail to integrate and communicate sustainability in their vision, mission, and goals and there is an absence of dialogue on sustainability issues with their stakeholders. This is usually observed in the case of start-ups and manufacturing companies that are either in the early stage of growth or are not concerned about their stakeholders. Also, sustainability is not seen as a strategic tool. For example, though Exxon Mobil constantly claims to incorporate sustainable elements into their business in the future, its main source of revenue is still dependent on fossil fuels. They are ranked highly on global emissions and have failed to practice low-carbon approaches in their business.[17] They have not been able to communicate and implement their sustainability strategy to ensure business sustainability while

---

[15] https://fsc.org/en/newsfeed/ikea-creating-a-sustainable-world.
[16] https://www.adidas.co.nz/blog/361051-the-oceans-death-by-plastic.
[17] https://www.bloomberg.com/news/articles/2020-10-05/exxon-carbon-emissions-and-climate-leaked-plans-reveal-rising-co2-output.

their competitors in the oil and gas industry have developed a sustainable strategy for the future.

Opaque companies claim to be sustainable and tend to overcook their claims on sustainable achievements. The main intention is to look 'Green' and market the same, rather than incorporating the same into their business. They fail to achieve on-ground impact but *'Green Wash'* their activities to build a positive public image to fuel their revenues. Some good examples for the same are the oil and gas companies. These companies have been historically degrading the environment, but are currently marketing sustainable practices to gain customer attention. A detail of sustainability greenwashing has been given in the next section.

Sustainability Greenwashing
In the current environmental scenario, greening the business is becoming a lucrative strategy to attract eco-lovers and catch their sentiments to grow more and earn profits. A report by Nielsen says that 66% of consumers worldwide are ready to spend more for buying products from sustainable brands. However, TerraChoice, an Environmental Marketing company says that in the name of sustainability almost 98% of so-called green/sustainable products are greenwashed.[18]

Environmentalist Jay Westervelt originally used 'greenwashing' as a terminology in 1986 in regard to the hotel industry to encourage green ideas through more than once usage of the towels to save resources.[19] Subsequently, the term was frequently being exploited by companies to showcase their false claims through misleading advertisements to propagate that they are undertaking greater responsibility for the environment than they truly are. This tendency of greenwashing has started hitting the roof in recent years with the incremental growth of communication tools when more and more companies try to hide their unsatisfactory environmental performance through positively projected communication (Delmas & Burbano, 2011). The authors further suggest a typology of companies looking at their sustainability performance and the way they

---

[18] https://www.feedough.com/what-is-greenwashing-types-examples/.

[19] https://www.downtoearth.org.in/blog/environment/the-deception-of-greenwashing-in-fast-fashion-75557.

communicate the same as '*greenwashing companies*' with positive communication and unsatisfactory sustainability performance; '*vocal green companies*' with positive sustainability communication backed with satisfactory sustainability performance; '*silent brown companies*' with no sustainability communication and unsatisfactory sustainability performance; and '*silent green companies*' with no sustainability communication and satisfactory sustainability performance. What is to be avoided at all costs by corporates is '*Sustainability Greenwashing*' where the primary focus is on '*appearing sustainable*' rather than '*being sustainable*' using communication tools which may cause major risk to shareholder value.[20]

In the year 2020, important companies like Shell, Johnson & Johnson, and IKEA are also among many others who have been accused of making dubious sustainability claims for which large-scale protests were made during COVID-19 pandemic as reported by eco-business. Some of the examples include a major Italian oil company Eni which was the first one to be prosecuted with a fine of €5 million for their misleading claim of palm oil-based diesel as '*green*' and environment friendly which European Union is going to phase out by 2030, due to possible threat of deforestation.[21] In another example, Shell inquired from people regarding their plan of action to reduce carbon footprint, through a poll. Through this poll they were trying to shift the responsibility for climate change to consumers for their emissions, as Alexandria Ocasio-Cortez tweeted. While responding to the debate about energy transition launched by Shell, congressmen responded that Shell already knew the dangers of emissions from fossil fuels since the 1990s. '*The audacity of Shell asking YOU what YOU'RE willing to do to reduce emissions. They're showing you RIGHT HERE how the suggestion that individual choices - not systems - are a main driver of climate change is a fossil fuel talking point*'.[22]

Similarly, while advertising their Windex Vinegar Ocean Plastic bottle, SC Johnson asserts that their window-cleaner bottle is made from recycled ocean plastic, as a part of their '*Help Seas Sparkle*' campaign.[23] The

---

[20] What is Sustainability Communication? Is it different than ESG Communication (the sustainability.io).

[21] https://www.eco-business.com/news/8-brands-called-out-for-greenwashing-in-2020/.

[22] https://www.nationofchange.org/2020/11/05/shell-oil-asked-what-you-are-willing-to-do-to-reduce-climate-change-enraging-advocates-and-activists/.

[23] https://www.windex.com/en-us/help-seas-sparkle.

'ocean plastic' claim suggests that the plastic was actually retrieved from the ocean which was actually not true as plastic retrieved from the ocean is not recyclable due to its degradation from sun and salt. Besides, the retrieval from the ocean is also unrealistic at such a large scale. Furniture giant IKEA's sustainability authorizations were also found fallacious as it was linked to using wood obtained from illegal logging in Ukraine.[24] Besides, its wood certification through Forest Stewardship Council, was also investigated by NGO Earthsight where it was found to be a body to greenwash the timber industry.[25]

Hence, a word of caution while sharing sustainability communication is a must keeping in mind the long-term interest of the company and its stakeholders. Deceiving the investors through a company's fake sustainability communication profile is risky and cannot last long. At some point of time, while conducting authenticity checks for validation, the truth on your sustainability transition will be revealed to stakeholders.[26]

Sustainability Communication Through Social Media
Increasing consciousness concerning sustainability and climate change issues among modern-day consumers and civil society has forced corporate organizations and government to undertake dedicated efforts to address as well as communicate genuinely about them. Social media has made it possible to communicate their sustainability initiatives in an appealing manner to create *'a positive brand image and reputation, as well as meaningful organization-stakeholder relationships'*. Lot of innovative and creative possibilities are there in social media communication concerning diversification of sustainability communication through stakeholders' engagement. Presenting the credible sustainability content through creative communication (style and format) can create a lot of interest among target audience (Dawkins, 2004) which can be further reinforced through inclusion of emotional content (Hartman et al., 2007) presented through interactive visual media (Khan et al., 2019). With the

---

[24] https://news.mongabay.com/2020/06/ikea-using-illegally-sourced-wood-from-ukraine-campaigners-say/.

[25] https://fsc.org/en/newsfeed/fsc-statement-on-earthsight-report-2020.

[26] The examples of companies given in previous pages to reflect sustainability communication including greenwashing are taken from different media reports and are merely for the purpose of explaining the concepts. Author has no intention to critically analyse the stated claims in favour of or against any of the company.

possibility of receiving direct feedback through its stakeholders, social media provides an added advantage over other forms of sustainability communication to companies (Basri & Siam, 2019). Through their social media pages, companies try to use creative tools to actively engage with their customers and convey their sustainability agendas focusing on how they are integrating ecological and social attributes to improve the brand perception (Richardson et al., 2020). Hoffmann, C., & Weithaler, L. (2015)[27] remarks... 'Interactive storytelling was found to be an efficient communicational tool to boost brand engagement among consumers in an era of vast information overload'.

The scope of a social media environment is vast, allowing people to share their real-time experiences and videos in the form of stories which consumers can relate with. Hence, storytelling can convey the message of sustainability which will be credible and trustworthy for the stakeholders and will keep the communication moving in social networking sites, thus having long-term impact in maintaining the chain of sustainability communication. Social media is steadily making people empowered with the sense that their concerns are significant for the companies and they can make a difference in the way companies behave as far as sustainability issues are concerned. The same is being well recognized by the companies too to behave and communicate in a responsible manner.[28]

Sustainable Consumption Communication—A Psychological Framework

Solution to most of the environmental problems and sustainability issues including sustainable consumption is closely aligned to human psychology and behavioural practices, ... *'the ways in which people think, interact, and behave'* (Clayton et al., 2016). For understanding the dynamics of this phenomenon, psychological theories can be suitably applied but concerning interventions in human behaviour for addressing sustainability challenges, integrating behavioural science with principles of communication is warranted. Communication is meant for people, the receivers of sustainability messages, hence, behavioural and psychological perspectives

---

[27] http://www.brandba.se/blog/2015/10/28/sustainability-communication-in-the-social-media-environment.

[28] https://www.theguardian.com/sustainable-business/communicating-sustainability-social-media-storytelling.

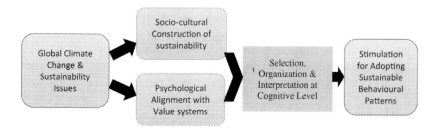

Fig. 3.5 A psychological framework of sustainability communication

play a very significant role in sustainability communication. Psychological framework of sustainability communication can be reflected as given below (Fig. 3.5).

Psychologically speaking, whenever we come across any climate change/sustainability issue, we operate at our cognitive level to see whether the issue is important enough or worth drawing our attention. This process is facilitated by social construction of issues/messages as well as their psychological alignment with existing value systems. Subsequently, we organize the associated cues in our mind based on experiences and draw interpretations. It is comprehended through both direct and mediated modes of communication. Hence, systematically designed communication is an imperative tool to inspire possible adoption of sustainable behavioural patterns (Kruse, 2011).

Richardson et al. (2020) explain sustainable consumption as a behavioural pattern that reduces the bearing on the natural world encompassing different behavioural facets like consumption behaviours, production practices as well as technological innovations. It is basically looking forward to behavioural transformation/shift in order to move from unsustainable to sustainable behavioural practices. Although there are a range of challenges in this process including habits, past experiences, long-held attitudes and behaviours, social-psychological research methodology and practices have the potential to help in understanding and explaining the challenges of sustainable consumption research especially pertaining to behavioural practices (Lim, 2017). A phenomenal piece of work in the field of environment–behaviour interrelationship has been undertaken to establish the application of applied psychology interventions in promoting '*environmentally significant behaviours*' and corresponding '*psychologically significant responses*' (Gifford et al., 2011; Swim et al., 2009).

For example, there is a communication regarding limited use of consumer goods (for instance-Mobile phone) and replacing it only when it becomes non-functional. At the very first level, this idea should cognitively appeal to you as something significant to deserve your attention. The first thing which comes in mind is what will people around me think if I am overusing an old mobile?; Whether and how much their thinking matters to me?; What is the gain/loss by replacing or not replacing functional mobile phone?; How justified it is to spend resources on that even if I have them?; What can be the justified use of discarded mobile phone if I choose to buy new one? What sustainability values have I been inculcated in and how do I feel embracing those values? Based on mental churning of associated thoughts at cognitive and affective level, the decision to act, i.e., replacing or not replacing a functional mobile phone is taken.

Similar psychological framework operates for buying any new consumer product and values for sustainability play a significant role while taking such decisions. Beattie and McGuire (2014) reported that there exists a gap between sustainability values and actual behaviour of people while making sustainable product choices/other pro-environmental decisions. Explicitly reflected attitudes are ruled by conscious mind to reveal what appears responsible to others (like sustainable consumptive practices) and latent attitudes are governed by subconscious mind which pushes people to act based on their long-held implicit behavioural tendencies. Empirical research shows significant differences between individuals' reflected values for sustainability and actual sustainability behaviours and most of the times implicit behavioural tendencies dominate action. Hence, the content and process of communicating sustainable consumption should consciously take into cognizance the underlying behavioural tendencies of people in order to minimize the values-action gap and produce environmentally desirable results.

Climate Change Communication

There is a symbiotic relationship between human beings and the natural environment in which they inhabit and extract various ecosystem services in order to assure their well-being (Leviston et al., 2018). 'Human actions are estimated to be causing the climate of the planet to change 170 times faster than natural forces'. states Global Catastrophic Risks Report

2020.[29] Further, Global Risks Report 2021 also says 'Among the highest likelihood risks of the next ten years are extreme weather, climate action failure and human-led environmental damage'.[30] Looking at the series of such reports on climate risks, there is strong evidence to proclaim that climate change is very much part of a public agenda. The objective of Climate Change Communication (C3) is to upkeep sustainable development and limit the adverse impact of climate change through public partnership and engagement. However, global media reporting of climate change is gradually declining which can be interpreted as early signs of 'climate fatigue'. It is making people, especially the younger generation, inert and less concerned to be cognitively aware of the phenomenon of climate change and their possible role in this regard (Wibeck, 2014).

One of the primary reasons for this inertia is the paradox that scientists communicate climate change as a scientific discipline and warrant technological interventions to mitigate its impacts. However, its far-reaching effects and impending approaches to mitigate them have pertinence for diverse population groups across the globe. Hence, communication strategies have to be customized to suit the specific groups of people in different regions and nations. Badullovich et al. (2020) mapped 281 research and bibliographic studies on climate change. About half of the frames were oriented towards science, economy, and environment followed by health issues, climate-linked disasters, and environmental ethics. Another significant finding was that more than half of the studies represented just one country, the United States, though C3 is quite sensitive to the respective country and socio-political context. A need for reflexive cross-country research on C3 was recommended with a focus on 'under-researched nations' to have its far-reaching impact.

Climate Change Communication (C3) can partly be held responsible for the visible triviality about or disowning of climate change and its undesirable consequences as well as anthropogenic causes by the public along with the necessity to mitigate or adapt with it. Besides that, Druckman and McGrath (2019), proclaim that C3 efficacy is dependent on intent and motivation of target audience. The expectations of the audience may be accuracy-driven or reiterative/belief-confirmation

---

[29] https://issuu.com/globalchallengesfoundation/docs/global_catastrophic_risks_2020_annual_report_web_v.

[30] http://www3.weforum.org/docs/WEF_The_Global_Risks_Report_2021.pdf.

driven which authors labelled as 'Accuracy motivation' vs. 'Directional motivation'. While accuracy-driven audience can be satisfied with content communicated on the bases of scientific consensus, belief–confirmation driven audience looks for the content which substantiates their long-held attitudes and beliefs about climate change. They find factual and scientifically-driven data having low credibility because their goal is 'belief–confirmation and identity protection'. So their intent is to question the expertise of scientists and credibility of the source of information, to disconfirm it.

Application of Fischhoff (1995)'s risk communication model projects that C3 promotes 'stress persuasion' instead of 'social movement mobilization'. Effective C3 does not necessarily need a varied set of strategies and self-reflection by communication agents. The quest for revealing limits to C3 is desirable (Johnson, 2012). The process of C3, as given in the following model, reflects that communication starts much before any word is spoken. The content as well as process of C3 are equally important and substantiate each other. The willingness to initiate conversation in itself transmits the message which is further augmented by one's deductions about content of communication. Hence, getting the authenticated factual data for climate change is the first step towards C3 otherwise, it may go at the wrong foot creating a scenario of trust-deficit (Fig. 3.6).

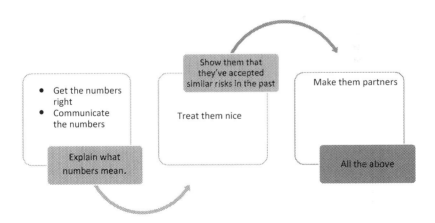

Fig. 3.6 The process of climate change communication (Adapted from Fischhoff [1995])

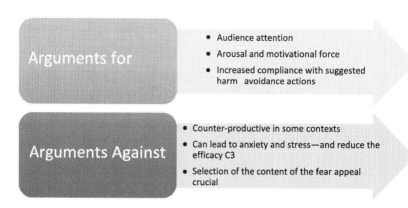

Fig. 3.7 Arguments for/against fear appeal for C3

Further to that, communicating the facts and figures which are backed up by the explanation which suits the cognitive needs of the audience is pertinent to proceed further. Once the audience is with you, rather than creating a fear appeal, they should be reassured that similar risks at different scales they have already come across. It will keep them motivated on the solution-centric plane rather than being trapped in the cycle of negative emotions, anxiety, and learned helplessness. According to this model, treating the audience nicely is very important in C3 so that they can subsequently be made partners in the process of initiating collaborative climate action. In this manner an empathic, psychologically oriented persuasive but collaborative strategy for C3 can be planned focusing on a broader set of values and taking into consideration diverse cultural contexts. In a study on binding communication on youth, a movie on climate change was shown which was followed by a pledge (persuasive action) to combat climate change which proved to be quite effective. Hence, communication followed by commitment can be one of the effective strategies for C3. Pairing of persuasive message/visuals with possible personal action concerning climate change mitigation can derive effective results which can have long-term impact in promoting pro-climate attitudes and behaviour (Parant et al., 2017) (Fig. 3.7).

C3 Through Fear Appeal—An Unanswered Debate
The above viewpoint is strongly supported by climate change researchers and communicators who believe that the coercive convincing to create

fear appeal is ineffective as well as incongruous to motivate climate change adaptation/action through public engagement. On the contrary, the drive reduction model (Higbee, 1969), already explained in previous chapter, states that fear appeal helps to increase level of arousal, which people are inherently motivated to reduce, to maintain homeostasis. Hence, communication during the state of high emotional arousal is supposed to be effective. Limited research supports this stream of thought but with inadequate empirical confirmation (Hartman, 2016). 'What grabs attention (dire predictions, extreme consequences) is often not what empowers action' (Moser & Dilling, 2007). The emotion of fear can act as a strong 'motivator as well as de-motivator of action'. Therefore, 'playing with emotional appeals to create urgency is like playing with fire' (Moser, 2007). There is a strong need for caution to select and use the content and process of C3 to obtain desired results as there are studies which reflect how creation of threat can be counterproductive too (Fritsche & Häfner, 2012). Hence, old models of communicating climate change through fear appeals may not always survive for long especially to convince the millennials. There are arguments in favour of as well as against fear appeals as reflected by Reser and Bradley (2017).

As arguments against fear appeal are many, it can be recommended that content and pictorial presentation in visual media, focusing on suggestive actions to be taken concerning climate change mitigation/adaptation, enthuse hope among individuals and they feel motivated to actively support climate change policies of the government. However, the reverse is commonly the case where, disregarding the focus on prospective individual actions, the media generally shows stories and imagery which depict adverse impacts of climate change in the form of frightening landscapes, taking people away from climate solutions (Feldman & Hart, 2018). Harvey et al. (2012) also pointed out that top-down approaches to C3 focusing on futuristic climatic scenarios and adverse climate events are losing ground. It generates fear and anxiety which the human psyche wants to avoid to overcome the negative emotional arousal (Higbee, 1969). Hence, an alienation from climate change related news and imagery is likely to develop, leading to possible public disengagement and the very objective of C3 gets diluted. Mann et al. (2017), also remarks that '…pessimistic coverage depresses and demoralizes the public into further inaction'. It is unarguably true that emotion-based messages can play a persuasive role in short-term climate change engagement but there is a need for utmost caution while using them in C3 as 'getting the affective science right may have significant benefits, but getting it wrong also has the potential for producing significant harm' or being counterproductive as aptly remarked by Chapman et al. (2017). Hence, public acceptability plays a crucial role in ensuring the efficacy of climate change mitigation policies (Ščasný et al., 2017) and the shift is now towards more process oriented approaches (Harvey et al., 2012).

Barriers to C3
With the increasing scholarly interest in C3 there is a growing concern regarding communicators to resolve uncertainties and controversies to build some consensus. Many times it acts as a barrier to efficacy of communicated content as people do not necessarily long for resolving their conflicting ideas and propositions concerning climate change. Moser and Dilling (2007) have emphatically demonstrated that limited efficacy of C3 can be attributed to four fallacious assumptions of climate change communicators:

- *Information Deficit Approach*—Limited information and understanding of climate change leads to constrained community engagement and resulting limited climate action (Sturgis & Allum, 2004). Hence, improving upon the content of C3 can lead to desirable climate appropriate behaviour.
- *Motivation by Fear Approach*—Catastrophic experiences of the past as a result of human failure to act appropriately can also motivate people for desirable climate action (Hartman, 2016).
- *Scientific Framing Approach*—Scientific presentation of climate change scenarios with facts and figures generated through evidence based research can move common people to action.
- *Media based Mobilization*—Public can be effectively mobilized for climate action through mass media due to its wider reach and strength. People trust what is presented in the media, as scientifically proven and take appropriate action.

Limited efficacy of above approaches can be explained through some of the barriers to C3. Examples of such pre-existing mental models can be like 'Industrial pollution leads to climate change'. Hence, individual energy consumption patterns are wilfully ignored by people, no matter how concrete and substantial information is provided to them through mass media. This barrier cannot be compensated through just better information, enhanced knowledge or effective modes of communication alone. There is a need for exploring behavioural change literature to devise effective strategies for behavioural transformation. This aspect is being dealt with in the forthcoming chapter on behavioural transformation for pro climate action.

Successful C3 requires identification of beliefs and values of people and exploiting them meaningfully to generate communication content supplemented by an effective process of communication. Integration of ICT-based visualizations and social media based communication models and innovative approach for their implementation within non-formal education may be what the twenty-first century is looking forward to. As the area of C3 is growing, empirical research on communication strategies and community engagement are being conducted in diverse regions and contexts by climate change communicators. However, they could enhance their engagement with environmental educators who have much wider professional experience of spreading awareness about environmental issues (Wibeck, 2014).

Pearce et al. (2015), remark that a better strategy to embrace the controversy and while engaging with the audience try to look for some democratically derived solution rather than being at the loggers head to counter and justify the communicated content. A pluralistic understanding of target audience with diverse beliefs and thought processes operating at cognitive and affective level and an inclusive approach of communicators (Harvey et al., 2012) taking into consideration the engagement of indigenous communities, who are likely to be most affected by climate change impacts (Fernández-Llamazares et al., 2015), can potentially counter barriers to C3.

COVID-19, Climate Change, and Sustainability Communication

COVID-19 has taken a serious toll on humanity as a whole but at the same time it is also being considered as a 'learning experience to deal with environmental sustainability and climate change'. It has also given the chance to look for the 'an opportunity to advance the climate agenda… and sustainable solutions in production and consumption' (Kakderi et al., 2021). Besides, it has posed severe challenges to experiential and outdoor environmental education programmes in almost all parts of the world. The existence of a virtual world of learning and communication can help realize sustainability and climate change related cognitive learning outcomes but affective learning and real-time experiential learning suffered a major setback.

In a multi-author analysis by Quay et al. (2020), Sandy Allen-Craig makes a very apt statement 'our post COVID 19 challenge: to ensure that outdoor and environmental education is not relegated to the virtual world, disconnected from mountains and forests, the natural tangible places where palpable learning can occur, where the mountains can speak for themselves'. Subsequently, Morten Asfeldt remarks that missing out on real-time learning through expeditions and outdoor learning in social environments can have adverse impacts on well-being and mental health, especially among youth.[31] Further, Simon Beames and others opined that 'Corona pandemic has amplified existing patterns of inequalities in use of the natural outdoor spaces. ...certain socio-cultural factors have enabled or constrained the ability of specific populations to access nearby natural spaces, and the health benefits that come with them'.[32] More so, it has been observed that communicators and educators have started over-communicating with heightened levels of empathy and compassion to better connect with virtual audiences. However, inequalities in availability of digital infrastructure are leaving the indigenous communities of developing nations far behind as far as environmental and sustainability education is concerned. Hence, there is an urgent need to educate cross-cultural populations with information pertaining to the '*ecological and behavioural drivers of coronavirus infection emergence and spread*' (Barouki et al., 2021).

There is an intricate connection between climate change, sustainability, and COVID-19/other pandemics, although there is no direct supporting data to substantiate this. However, many of the fundamental causes of climate change also increase pandemics. For example, deforestation causes climate change and also leads to animal migration. As a result, they come in contact with animals, they are not generally exposed to, resulting in disease transmission among them which can be transmitted to human beings too.[33] Similarly, many other natural habitats and ecosystems are

---

[31] https://link.springer.com/article/10.1007%2Fs42322-020-00059-2.

[32] https://link.springer.com/article/10.1007%2Fs42322-020-00059-2.

[33] https://www.hsph.harvard.edu/c-change/subtopics/coronavirus-and-climate-change/.

undesirably altered due to changing climate, impacting the species living within that ecosystem. A study discovered that changing climatic conditions created a conducive environment for some of the bat species to allow the emergence of novel coronaviruses.[34] Hence, our very action leads to a reaction in the natural environment which can be disastrous now or in the times to come. It can be especially true for the marginalized classes and disadvantaged groups, having limited adaptive capacity, whether or not the crisis is created out of climate change or pandemic. As aptly remarked, 'Both COVID-19 and the climate crisis have exposed the fact that the poorest and most marginalised people in society, such as migrants and refugee populations, are always the most vulnerable to shocks'.[35]

Climate communication concerning COVID-19 should be strategized in a manner that people can understand at a relatable psychological distance. Communication of such relevant content backed up with facts and figures will be useful, if communicated through unpretentious messaging which people can relate with and trust. Abstraction leads to attentional crowding and to clear the clutter which people might experience, many times relevant issues which are not backed up by personally relatable messages, are consciously left out (Hochachka, 2020).

Klenert et al. (2020) draw teachings from COVID-19 which can be adopted for moving ahead with climate change and sustainability linked crises. The higher cost of delay in communicating the threat to people appropriately and failure to take time-bound action is the first learning from COVID-19 crisis. The countries taking quick restrictive measures to curb the spread, were the ones who could limit the devastating impact to some extent. This quick action was either the result of previous experience with epidemic outbreaks or learning from the experience of others. The similar lessons are warranted concerning climate change also before it becomes too late and devastating impacts become publicly visible. Secondly, engagement with civil society and taking people on board is equally important which is possible only if understandable content of communication is transmitted through context

---

[34] https://www.news-medical.net/health/Climate-Change-and-COVID-19.aspx.
[35] https://www.thelancet.com/journals/lancet/article/PIIS0140-6736(20)32579-4/fulltext.

appropriate process/communication strategies. Campaigning for information and awareness generation concerning any crisis should be reframed in such a manner that the psychological distance to crisis is reduced and public perceptions are consistent with factual information with minimum psychological gap (Jones et al., 2017) which equally applies for climate change as well as pandemic induced crisis. Thirdly, inequalities can burden as well as worsen the situation. Both climatic changes and pandemic have the potential to disproportionately impact marginalized classes and disadvantaged groups.[36]

Another significant learning which has been drawn concerning this issue is that '*global problems necessitate global cooperation*' as narrated by (Klenert et al., 2020). He further establishes this statement with the support of cross-country examples and role of international bodies that 'COVID-19 demonstrates the need for these institutions to remain apolitical in giving evidence-based advice and setting policy' which is inclusive in nature and capable of extending holistic benefits to all parts of the world. It is followed by the final learning from the author pertaining to the above that no one can disregard the scientific advice, however scientific advice should be framed in a manner which balances and takes into consideration the inherent value judgements of respective countries so that it is relatable to people. Scientific evidence based policies and their implementation with civil society on board are required. Otherwise, the spread of misinformation can jeopardize the systemic implementation of any scientifically designed policy. Human beings are the key players and their affective/value based concerns should be cautiously manipulated in favour of scientific evidence to have a larger impact on crisis mitigation whether climate change or COVID-19.

Conclusion—Shaping of a Sustainable Society

With an increased scarcity of natural resources, and increasing awareness towards climate change, the modern societies have been propelled to shift towards renewable energy and eco-friendly products. In order to orient the present generation to this kind of change and also to expedite the process, climate change communication (C3) is of utmost importance. Increasing number of global initiatives to achieve SDGs make it pertinent

---

[36] https://www.iza.org/publications/dp/13183.

to actively involve our society in working towards achieving these goals as well. If we aim to build a climate-conscious society, it is necessary to not just communicate information about climate change and sustainability, but it is also important to embed these values in our everyday lives.

Immediate actions concerning sustainability and climate change reporting by businesses are imperative. In addition, communicating to all the stakeholders about the company's commitment towards SDGs and our planet is also important in order to build a positive outlook for the brand. As demonstrated by successful businesses around the world, *'creating shared value'* for all the stakeholders through sustainability initiatives can help in attaining strategic advantage. However, it has been observed that foul practices have been used by several companies to project a false brand image via wrongful sustainability reporting and marketing. The modern society has been subjected to false marketing through *'greenwashing'* where companies have falsely claimed to be green and sustainable. This false propaganda is extremely undesirable and can lead to a negative long-term impact on the brand image.

As sustainability communications are targeted towards society, behavioural and psychological aspects of the same must be duly considered. Utilizing modern-day social media platforms to direct the communications can help in understanding real-time response and impact of the same. But possible adoption of these considerations by our society can mainly be influenced by aligning the campaigns with already existing value systems. It is important to bring about positive energy among individuals to embrace eco-friendly behaviour.

Dynamic relation between society and nature has been the key to the current climate change crisis. Through C3, the current generation can be made aware of its role in mitigating climate change. But considering the current socio-economic and global political situation, this matter needs to be handled with great care. A balance has to be maintained between *'number-driven communication'* and *'storytelling-based'* reporting to have a wider reach and influence. Inclusive communication will play a vital role in ensuring the engagement of vast varieties of communities on a global scale. It is pertinent to ensure that no one is left out, including indigenous communities who have been practicing traditional climate change combating techniques.

It is the time that we analyse and ponder upon how the current COVID-19 induced pandemic has provided us with an opportunity to re-access and plan our climate-related communication goals. It is vital to identify the changed values and beliefs in the society and target sustainability communication along those lines to achieve massive traction. Finally, it is also a time to invoke social and environmental responsibility within corporations and society as a whole via targeted sustainability and climate-related communications to bring about a positive change for our planet. Industry can play a very responsible role to promote climate change and sustainability education and communication under their Corporate Social Responsibility initiatives where it can serve as one of the significant contribution to SDGs, as emphasized in next chapter.

REFERENCES

Badullovich, N., Grant, W. J., & Colvin, R. M. (2020). Framing climate change for effective communication: A systematic map. *Environmental Research Letters, 15*(12), 123002. https://doi.org/10.1088/1748-9326/aba4c7

Baldassarre, F., & Campo, R. (2016). Sustainability as a marketing tool: To be or to appear to be? *Business Horizons, 59*(4), 421–429. https://doi.org/10.1016/j.bushor.2016.03.005

Barouki, R., Kogevinas, M., Audouze, K., Belesova, K., Bergman, A., Birnbaum, L., Boekhold, S., Denys, S., Desseille, C., & Drakvik, E. (2021). The COVID-19 pandemic and global environmental change: Emerging research needs. *Environment International, 146*, 106272.

Basri, W. S. M., & Siam, M. R. (2019). Social media and corporate communication antecedents of SME sustainability performance: A conceptual framework for SMEs of Arab world. *Journal of Economic and Administrative Sciences*.

Beattie, G., & McGuire, L. (2014). The psychology of sustainable consumption. In *Sustainable consumption* (pp. 175–195). Oxford University Press.

Bilharz, M., & Schmitt, K. (2011). Going big with big matters: The key points approach to sustainable consumption. *GAIA-Ecological Perspectives for Science and Society, 20*(4), 232–235.

Brand, K.-W. (2006). *Die neue Dynamik des Bio-Markts: Folgen der Agrarwende im Bereich Landwirtschaft, Verarbeitung, Handel, Konsum und Ernährungskommunikation; Ergebnisband 1*. ökom-Verlag.

Chapman, D. A., Lickel, B., & Markowitz, E. M. (2017). Reassessing emotion in climate change communication. *Nature Climate Change, 7*(12), 850–852.

Clayton, S., Devine-Wright, P., Swim, J., Bonnes, M., Steg, L., Whitmarsh, L., & Carrico, A. (2016). Expanding the role for psychology in addressing environmental challenges. *American Psychologist, 71*(3), 199–215. https://doi.org/10.1037/a0039482

Dawkins, J. (2004). Corporate responsibility: The communication challenge. *Journal of Communication Management*.

Delmas, M. A., & Burbano, V. C. (2011). The drivers of Greenwashing. *California Management Review, 54*(1), 64–87. https://doi.org/10.1525/cmr.2011.54.1.64

Doorley, J., & Garcia, H. F. (2015). *Reputation management: The key to successful public relations and corporate communication*. Routledge.

Druckman, J. N., & McGrath, M. C. (2019). The evidence for motivated reasoning in climate change preference formation. *Nature Climate Change, 9*(2), 111–119. https://doi.org/10.1038/s41558-018-0360-1

Feldman, L., & Hart, P. S. (2018). Is there any hope? How climate change news imagery and text influence audience emotions and support for climate mitigation policies: Is there any hope? *Risk Analysis, 38*(3), 585–602. https://doi.org/10.1111/risa.12868

Fernández-Llamazares, Á., Méndez-López, M. E., Díaz-Reviriego, I., McBride, M. F., Pyhälä, A., Rosell-Melé, A., & Reyes-García, V. (2015). Links between media communication and local perceptions of climate change in an indigenous society. *Climatic Change, 131*(2), 307–320. https://doi.org/10.1007/s10584-015-1381-7

Fischhoff, B. (1995). Risk perception and communication unplugged: Twenty years of process 1. *Risk Analysis, 15*(2), 137–145.

Fritsche, I., & Häfner, K. (2012). The malicious effects of existential threat on motivation to protect the natural environment and the role of environmental identity as a moderator. *Environment and Behavior, 44*(4), 570–590.

Galpin, T., Whitttington, J. L., & Bell, G. (2015). Is your sustainability strategy sustainable? Creating a culture of sustainability. *Corporate Governance*.

Genç, A., Patarroyo, J., Sancho-Parramon, J., Bastús, N. G., Puntes, V., & Arbiol, J. (2017). Hollow metal nanostructures for enhanced plasmonics: Synthesis, local plasmonic properties and applications. *Nanophotonics, 6*(1), 193–213.

Gifford, R., Steg, L., & Reser, J. P. (2011). Environmental psychology. In P. R. Martin, F. M. Cheung, M. C. Knowles, M. Kyrios, J. B. Overmier, & J. M. Prieto (Eds.), *IAAP handbook of applied psychology* (pp. 440–470). Wiley-Blackwell. https://doi.org/10.1002/9781444395150.ch18

Godemann, J., & Michelsen, G. (2011). Sustainability communication–an introduction. In *Sustainability communication* (pp. 3–11). Springer.

Hartman, L. P., Rubin, R. S., & Dhanda, K. K. (2007). The communication of corporate social responsibility: United States and European Union multinational corporations. *Journal of Business Ethics, 74*(4), 373–389.

Hartman, S. (2016). Climate change, public engagement, and integrated environmental humanities. *Teaching Climate Change in the Humanities*, 67–75.

Harvey, B., Carlile, L., Ensor, J., Garside, B., & Patterson, Z. (2012). Understanding context in learning-centred approaches to climate change communication. *IDS Bulletin, 43*(5), 31–37.

Hespenheide, E., Pavlovsky, K., & McElroy, M. (2010). Accounting for sustainability performance: Organizations that manage and measure sustainability effectively could see benefits to their brand and shareholder engagement and retention as well as to their financial bottom line. *Financial Executive, 26*(2), 52–57.

Higbee, K. L. (1969). Fifteen years of fear arousal: Research on threat appeals: 1953–1968. *Psychological Bulletin, 72*(6), 426.

Hochachka, G. (2020). Unearthing insights for climate change response in the midst of the COVID-19 pandemic. *Global Sustainability, 3*, e33. https://doi.org/10.1017/sus.2020.27

Hoffmann, C., & Weithaler, L. (2015). *Building brand reputation in the digital age: Identifying effective brand communication to win the moment of truth online.*

Johnson, B. B. (2012). Climate change communication: A provocative inquiry into motives, meanings, and means: Perspective. *Risk Analysis, 32*(6), 973–991. https://doi.org/10.1111/j.1539-6924.2011.01731.x

Jones, C., Hine, D. W., & Marks, A. D. (2017). The future is now: Reducing psychological distance to increase public engagement with climate change. *Risk Analysis, 37*(2), 331–341.

Kakderi, C., Komninos, N., Panori, A., & Oikonomaki, E. (2021). Next city: Learning from cities during COVID-19 to Tackle climate change. *Sustainability, 13*(6), 3158.

Khan, A. A., Wang, M. Z., Ehsan, S., Nurunnabi, M., & Hashmi, M. H. (2019). Linking sustainability-oriented marketing to social media and web atmospheric cues. *Sustainability, 11*(9), 2663.

Klenert, D., Funke, F., Mattauch, L., & O'Callaghan, B. (2020). Five lessons from COVID-19 for advancing climate change mitigation. *Environmental and Resource Economics, 76*(4), 751–778.

Kruse, L. (2011). Psychological aspects of sustainability communication. In *Sustainability communication* (pp. 69–77). Springer.

Leviston, Z., Walker, I., Green, M., & Price, J. (2018). Linkages between ecosystem services and human wellbeing: A Nexus Webs approach. *Ecological Indicators, 93*, 658–668. https://doi.org/10.1016/j.ecolind.2018.05.052

Lim, W. M. (2017). Inside the sustainable consumption theoretical toolbox: Critical concepts for sustainability, consumption, and marketing. *Journal of Business Research, 78*, 69–80.

López, B., & Monfort, A. (2017). Creating shared value in the context of sustainability: The communication strategy of MNCs. *Corporate Governance and Strategic Decision Making*, 119–135.

Luhmann, N. (1997). Globalization or world society: How to conceive of modern society? *International Review of Sociology, 7*(1), 67–79.

Mann, M. E., Hassol, S. J., & Toles, T. (2017). Doomsday scenarios are as harmful as climate change denial. *The Washington Post*, 12.

Minott, D., Ferguson, T., & Minott, G. (2019). Critical thinking and sustainable development. In W. Leal Filho (Ed.), *Encyclopedia of sustainability in higher education* (pp. 1–6). Springer International Publishing. https://doi.org/10.1007/978-3-319-63951-2_529-1

Moser, S. C. (2007). *More bad news: The risk of neglecting emotional responses to climate change information.*

Moser, S. C., & Dilling, L. (2007). Toward the social tipping point: Creating a climate for change. In S. C. Moser & L. Dilling (Eds.), *Creating a climate for change* (pp. 491–516). Cambridge University Press. https://doi.org/10.1017/CBO9780511535871.035

Newig, J. (2011). Climate change as an element of sustainability communication. In *Sustainability communication* (pp. 119–128). Springer.

Newig, J., Schulz, D., Fischer, D., Hetze, K., Laws, N., Lüdecke, G., & Rieckmann, M. (2013). Communication regarding sustainability: Conceptual perspectives and exploration of societal subsystems. *Sustainability, 5*(7), 2976–2990.

Parant, A., Pascual, A., Jugel, M., Kerroume, M., Felonneau, M.-L., & Guéguen, N. (2017). Raising students awareness to climate change: An illustration with binding communication. *Environment and Behavior, 49*(3), 339–353. https://doi.org/10.1177/0013916516629191

Pearce, W., Brown, B., Nerlich, B., & Koteyko, N. (2015). Communicating climate change: Conduits, content, and consensus. *WIREs Climate Change, 6*(6), 613–626. https://doi.org/10.1002/wcc.366

Quay, J., Gray, T., Thomas, G., Allen-Craig, S., Asfeldt, M., Andkjaer, S., Beames, S., Cosgriff, M., Dyment, J., Higgins, P., Ho, S., Leather, M., Mitten, D., Morse, M., Neill, J., North, C., Passy, R., Pedersen-Gurholt, K., Polley, S., ... Foley, D. (2020). What future/s for outdoor and environmental education in a world that has contended with COVID-19? *Journal of Outdoor and Environmental Education, 23*(2), 93–117. https://doi.org/10.1007/s42322-020-00059-2

Reser, J. P., & Bradley, G. L. (2017). Fear appeals in climate change communication. In *Oxford research encyclopedia of climate science*.

Richardson, L. M., Ginn, J., Prosser, A. M. B., Fernando, J. W., & Judge, M. (2020). Improving research on the psychology of sustainable consumption: Some considerations from an early career perspective. *Journal of Social Issues, 76*(1), 150–163. https://doi.org/10.1111/josi.12373

Ščasný, M., Zvěřinová, I., Czajkowski, M., Kyselá, E., & Zagórska, K. (2017). Public acceptability of climate change mitigation policies: A discrete choice experiment. *Climate Policy, 17*(sup1), S111–S130.

Sturgis, P., & Allum, N. (2004). Science in society: Re-evaluating the deficit model of public attitudes. *Public Understanding of Science, 13*(1), 55–74.

Swim, J., Clayton, S., Doherty, T., Gifford, R., Howard, G., Reser, J., Stern, P., & Weber, E. (2009). Psychology and global climate change: Addressing a multi-faceted phenomenon and set of challenges: A report by the American Psychological Association's task force on the interface between psychology and global climate change. American Psychological Association.

von Kutzschenbach, M., & Brønn, C. (2006). Communicating sustainable development initiatives: Applying co-orientation to forest management certification. *Journal of Communication Management, 10*(3), 304–322. https://doi.org/10.1108/13632540610681185

Wibeck, V. (2014). Enhancing learning, communication and public engagement about climate change—Some lessons from recent literature. *Environmental Education Research, 20*(3), 387–411. https://doi.org/10.1080/13504622.2013.812720

CHAPTER 4

Frugality and Innovation for Sustainability

INTRODUCTION

Globalization has geared up the ever-increasing tendencies of many of the developed nations to long for mass-consumption practices. These tendencies are more prominent in some of the societies and people, who are more advantaged over others in terms of resource availability. It is true that the corporate world primarily survives on such segments of society but it takes a heavy toll on the very concept of sustainability which has now been internationally recognized as a part of responsible business practices. 'He who will not economize will have to agonize', says Confucius (Analects).[1] Sustainable behaviour includes within itself the inclusive approach to extend compassionate care to inhabitants of planet earth besides mere conservation of living and non-living substrates (Corral-Verdugo et al., 2013). Therefore, 'Ensuring responsible consumption and production' by 2030 is one of the Sustainable Development Goals (SDG 12). As a part of targets for SDG 12, global population must have adequate information and awareness about sustainable production and consumption practices and should be motivated to adopt lifestyles which ensure harmonious relationship with nature.[2] Much before SDGs,

---

[1] Coenn, D. (2014). *Confucius: His Words*. BookRix.
[2] https://www.un.org/development/desa/disabilities/envision2030-goal12.html.

© The Author(s), under exclusive license to Springer Nature Singapore Pte Ltd. 2022
P. Rishi, *Managing Climate Change and Sustainability through Behavioural Transformation*, Sustainable Development Goals Series,
https://Doi.org/10.1007/978-981-16-8519-4_4

79

Agenda 21 of the UN's action plan for sustainable development also recommended promotion of positive attitudes and behavioural practices towards responsible and sustainable consumptive practices on the part of government and non-governmental organizations 'through education, public awareness programmes and other means [...]' (United Nations Environment Programme, 1992, sec. 4.3).[3]

Ensuring equity at social, ecological, and economic fronts is what sustainability is looking forward to. Sustainable solutions to foremost global challenges warrant equitable treatment of global inhabitants. 'Hunger, poverty, social injustice, and general lack of resources that afflict billions of people, are all important aspects to consider when ideating for large-scale sustainability solutions' (Basu et al., 2013). The time has come when we have to rise above the ego-centric tendencies of *me and mine* and move towards the behavioural practices which are morally correct and altruistic as well as inclusive in nature described as *we all*, encompassing all segments of society but especially taking into consideration the bottom of the pyramid.

The increasing disparities and inequalities worldwide, in terms of resource availability, explain the already given quote of Confucius in a way that because a limited number of privileged class is not economizing their lifestyle and behavioural practices, the larger segment of human society is forced to agonize (Roiland, 2016), i.e. living a life that is constrained in terms of sustainable resource availability. We have already observed the disastrous outcomes of unmindful non-renewable energy use and buying behaviour of consumer products as an outcome of our unsustainable behavioural practices. Unfortunately, the societies which are most affected by these consequences are the least contributors, raising questions about how long we can continue to adopt our behavioural practices which are creating miserable situations for our fellow inhabitants on planet earth. We all share a common responsibility for planet earth and a collective action is warranted towards reaching the targets of sustainable development goals on behalf of all the countries. The concept of Frugality having origin in Asian countries can partially answer to address this issue.

---

[3] United Nations Environment Programme (1992). *Agenda 21*. United Nations.

The Concept of Frugality

Frugality is described as 'a unidimensional consumer lifestyle trait characterized by the degree to which consumers are both restrained in acquiring and resourceful in using economic goods and services to achieve longer-term goals' (Lastovicka et al., 1999). Frugality is a central and distinctive feature of a sustainable lifestyle warranting reduced consumption and thereby reduced impact of behavioural practices on the *'availability and renewability of natural resources'* (De Young, 1996). As an antagonist to prevailing consumerism (Jackson, 2012), Frugal behavioural practices involve reduced day-to-day purchasing, consumption, and waste disposal patterns for the cause of ecological sustainability.

The concept of frugality also lays emphasis on universal good through minimalistic lifestyle and consumptive practices. Bouckaert et al. (2008) in their book on *'Frugality: Rebalancing Material and Spiritual Values in Economic Life'* explain how frugality is a source of collective contentment leading to sustainability through *'intergenerational justice'*. He further explains that frugality-based behavioural practices are likely to be linked with spiritual mindset leading to tranquillity and social peace, opening our mind to opt for rational choices in favour of ecological and social sustainability. In such a manner, preliminary connections between frugality and sustainability were established, allowing people to move towards holistic long-term happiness through balancing of material, social and spiritual needs. Nobel Laureate Kahneman, in his two system theory of decision-making, narrates that system two thinking gets activated when we are consciously adopting a particular behavioural practice (Kahneman, 2011) for the cause of society and environment. Frugality-based behavioural practices are likely to be more rational and reasonable being the outcome of thinking slow while consciously attending to all social and environmental clues before making any decision. On the other hand, system one thinking governs all fast and impulsive decisions which may not be thoughtful and aligned to social and environmental goals.

Besides, there are some other terminologies in practice which closely relate to frugality, like voluntary simplicity by Alexander (2011), which discards consumption-centric and materialistic lifestyle and promotes contentment through non-materialistic modes of living, having deeper connection with ecological sustainability and sustainable development. In Asia, Dayalbagh Eco-city of India is another example of voluntary simplistic lifestyle and behavioural practices. It has roots in spirituality, but through frugal innovations, it beautifully connects to sustainable development as described in detail in the subsequent sections of this chapter. Similar models of voluntary simplicity and moderation of consumer needs through frugal innovations were grossly appreciated in some parts of the German society too (Tiwari, 2017). Further, it was reported that millennials, who are otherwise considered as material-intensive, are also going for the minimalist lifestyle and practices (Radjou & Prabhu, 2015). Sandler (2009) also states that frugal behavioural practices are consciously conducive for sustainability. Japan, as a country, is a classic example of practicing the tradition of frugality and especially in recent times, the minimalist approach is increasingly being adapted and practiced by more and more people, especially the young generation. Taking inspiration from countries like Japan, citizens of all the countries should be educated and trained for frugality and its importance for individual as well as societal welfare of the present and future generations (Zwarthoed, 2015).

FRUGALITY ACROSS DIVERSE SOCIETIES AND CULTURES

'Frugality is founded on the principle that all riches have limits' remarks Edmund Burke.[4] However, the macro-level choices and decision-making for climate change and sustainability are primarily governed by the pre-existing social, cultural, and spiritual thought processes too, besides the principles of logic and rationality.

In the United States, consumerism and success are identified with each other. Japan's sacred treasures were television, air conditioning, and automobiles (Durning, 1991), however, there have been recent trends towards frugal behavioural practices in japan as described in previous section. On the other hand, the *'happy little kingdom'* of Bhutan co-creates happiness for everyone with very limited resources, which are just enough to live a decent life. The in-depth analysis of this case is given in the subsequent section. This indicates that frugality-based lifestyles and practices vary across diverse societies and cultures. Justinian's quote,

---

[4] https://ayietim.com/2016/02/15/edmund-burke-frugality-is-founded-on-the-princi pal-that-all-riches-have-limits/.

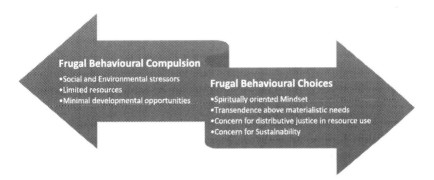

Fig. 4.1 Frugality—behavioural consumption vs. choice

'Frugality is the mother of all virtues'[5] very well explains how frugality connects with nature. It takes people on the path of virtues and sharing and caring, for contentment and global good which is over and above greed, unmindful resource exploitation and consumptive practices, taking people nowhere. Therefore, frugality can be explained as a *'behavioural choice'* vs. *'behavioural compulsion'*. Frugal behavioural choice is guided by spiritually oriented mindset and a desire for transcendence above materialistic needs while frugal behavioural compulsion is the outcome of social/environmental stressors. It has the potential to go for frugal innovations commonly known as *'Jugaad'* in Hindi language, adopted to compensate the resource-intensive needs in a simplistic and low-cost manner which will be described further (Fig. 4.1).

In a country like India, where spiritual and cultural ethos were quite strong at one point of time, the socio-psychological effects of over commercialization and globalization are becoming visible in major urban centres indicating the need for frugality-based behaviours. India is primarily an agriculture-based economy with 82% of small and marginal farmers with limited resources and developmental indices.[6] With a highly diverse Indian population (Chhokar, 2007), a major chunk is still living

---

[5] http://www.knowyourquotes.com/Frugality-Is-The-Mother-Of-All-Virtues-Justinian. html#:~:text=other%20human%20virtues.-,GIUSEPPE%20MAZZINI%20Humility%20is% 20the%20solid%20foundation%20of%20all%20virtues,is%20one%20thing%2C%20avarice% 20another.

[6] http://www.fao.org/india/fao-in-india/india-at-a-glance/en/.

in the rural areas or country ridges, far away from the availability and economically affording capacity of the consumer products, managing subsistence at the bottom of the pyramid. The result may be economically and developmentally dissatisfying, but at the sustainability and climate front, the carbon emission levels are quite low in comparison to other consumer-centric countries. The environmental appraisal of such areas may differ quite substantially from the point of view of local residents and the economic analysts with all rationality and logic. The Human Development reports also proclaim that the challenge of development is to improve the quality of life at the bottom of the pyramid. Now the question is, 'from which perspective'? Quality of life may have the social perspective, the psychological perspective, the environmental perspective, the economic perspective or even the spiritual perspective. In spite of having very low developmental indices, India ranks third in the world with most billionaires.[7] With ever-increasing disparity between rich and poor and a greater need for developmental infrastructure in rural areas, CSR was mandated for eligible companies under section 135 of companies' act 2013[8] which will be exclusively discussed in subsequent chapters. However, more than corporate social responsibility, there is a need for Individual Social Responsibility (ISR) too, where people are expected to be compassionate and caring for fellow citizens besides adopting frugal behavioural practices by choice so that under-privileged class and generations to come can have a fair share of their resources.

Frugality, Values, and Consumerism

While the world has progressed economically, technologically, or scientifically, human beings seem to have regressed despite their intellectual and materialistic progress for the sake of preserving the global climate. 'The greatest wealth is to live content with little' remarks Plato. As an outcome of the revolution of progress and consequent mass production and consumptive system, human beings are getting more and more mechanized, specialized, and standardized. Unprecedented techno-economic and scientific achievements are trapping people in acquisitive culture

---

[7] https://indianexpress.com/article/business/india-top-richest-billionaires-mukesh-ambani-gautam-adani-shiv-nadar-radhakishan-damani-uday-kotak-7264167/.

[8] https://www.mca.gov.in/Ministry/pdf/CompaniesAct2013.pdf.

intended for sensory gratification of lower level and selfish attitudes. A high standard of living is equated not with better 'thoughts' but with better consumer products and materialistic possessions. The great tragedy of the civilization is that it has improved the living standard of human beings but not the human beings themselves and at the same time caused great damage to climate and natural environment. Materialistic lifestyles in the west and population explosion in the east, especially in some of the Asian countries like China and India, have abnormally increased the demand for consumer goods. It is high time that we think of going back to the great Gandhian thought, '*Simple living and high thinking*'. Landes (2019)[9] comments that over engagement of larger population groups in frugal behaviour can lead to economic depression while absence of frugality behaviour can lead to '*impulsive - compulsive consumer behaviour*' (Shoham et al., 2017) and can have negative consequences on environment and sustainable behavioural practices (Parkins & Craig, 2011).

There has been a growing interest in frugality among researchers (Bove et al., 2009) and scholars, primarily due to increasing environmental concerns among today's much more aware consumers. Some of the modern management scientists believe that the very objective of frugality is to reduce consumption which can be counterproductive for the progressive business world. However, entrepreneurial companies are able to insightfully provide frugal products and services in the form of a successful business model too, adding meaningfully to their environmental, social, and economic value of business (Rosca et al., 2017). Therefore, the sustainability of production and consumption (SDG 12) needs serious attention on the part of companies as well as consumers, taking into consideration, the multiple perspectives.

The above analysis raises a few old but still very pertinent questions like—Are we heading towards a value system based and virtue loaded world full of joy, compassion, love, or concern? Or just a task-oriented world? (Nathawat, 1975). Frugality model is the combination of value based approach coupled with task orientation for the cause of the environment. It is also deeply rooted in some of the religious and spiritual practices as described below.

---

[9] https://www.consumerismcommentary.com/frugality-is-bad-for-the-economy/.

Frugal Traditions in Eastern Religions

The very concept of frugality has its origins in philosophy, being deeply rooted in some of the eastern and western religious practices. 'Deprive yourself on nothing necessary for your comfort, but live in an honourable simplicity and frugality', remarks John McDonough.[10] Further, Westacott (2016) in his book '*The Wisdom of Frugality*' explores and analyses the concept of simplicity as against materialism and individualism, and its role in creating a happier and healthier society from a philosophical perspective. Asceticism, austerity, and self-restriction are popularly practiced in Asian Buddhist traditions (Bouckaert et al., 2008). Many of the Asian societies and communities are prone to natural disasters, leaving them with limited resources and developmental opportunities. To face such environmentally stressful situations, religions like Jainism, Buddhism, Confucianism, and Daoism promote the importance of frugality as a part of socially and spiritually desirable lifestyle, to promote better resilience (Roiland, 2016). Many of the Indian people following spiritual paths decide to follow a frugal life living in simplicity away from the buzz of modern life and secluded in monasteries on the mountains. They show us the path of eternal happiness in complete harmony with nature. Many of the eastern religions like Jainism, Buddhism, Confucianism, and Taoism have been practicing strong frugal practices since centuries. Daoists were considered to be one of the foremost proponents of frugality, which is now popularly known as '*voluntary simplicity*'.[11] But how far the objective behind those behavioural practices was nature conservation or mere spiritual, needs an in-depth study of their religious texts.

One such religion, Jainism, probably the oldest one in the world (seventh century BC) also proposed the path of asceticism with a complete surrender and gratitude towards nature. Jainism recommends frugality and does not encourage the unconstrained accrual of affluence, as over-possessiveness and material attachment can lead to aberration of thought processes.[12] One of the sacred principles of Jainism is '*sanyama*' which means self-control or remaining satisfied with whatever minimum is available to oneself and leading a simple and self-restraint life full of

---

[10] https://impossiblyfrugal.com/frugal-quote-john-mcdonough/.

[11] http://urbanecohermit.blogspot.com/2010/10/dao-and-frugality.html.

[12] https://www.longfinance.net/news/pamphleteers/jains-wealth-and-ethics-lessons-godless-capitalism/.

sufficiency and frugality. It is recommended that if one is trapped in the idea of having more and more, there is no end to it. Hence, Jains are required to move from '*more and more*' to '*enough*!' It is recommended that there is an abundance of nature but if one tries to overpower and possess it, a kind of scarcity will be created as the desire to have more and more is unstoppable.[13] Contentment leads to abundance and hence, frugality-based lifestyle teaches us how to limit our needs for the cause of our own peace as well as for the cause of mother nature's sustainable existence. Similar practices were prevalent in Buddhist religion too.

Buddhism is known for its strong frugal traditions spanning across a very long history. Lord Buddha recommended frugality centuries ago emphasizing that nothing ceases in life. It only '*transforms, reinvents and reincarnates*',[14] which can be considered as one of the initial lessons for 3R (Reduce, Reuse, Recycle) of sustainable development. While advising a common man, Lord Buddha says that '*changing one's lifestyle and habits to avoid wasteful pursuits*' is what frugality is all about as mentioned in the *SigalovadaSutra* (DN 31).[15] Principles of Buddhism state that accumulating the wealth and attachment to that leads to desire to have more and more which is a constant source of dissatisfaction.

Looking at the trends of modern civilization, economic welfare of the society and the individuals and the organization of a socialistic pattern are the predominant objects of all social activity. But if one tries to find out whether economically better off individuals and communities are happier than those that are not so well placed, we would discover that they are not necessarily so. As against the materialistic and consumer-centric western world longing for possessiveness and associating it with well-being, Kaplan (1991) raises a very pertinent question 'Are billions of people of the developing world without modern consumer products and transportations systems, in poor psychological health?' A case of Bhutan, practicing Buddhism, and a case of Dayalbagh (a small township near Agra, the city of Tajmahal), a spiritual community following *Radhasoami* faith,[16] exhibit live examples of their strong frugal traditions coupled with sustainable development initiatives.

---

[13] https://www.resurgence.org/satish-kumar/articles/jain-religion.html.
[14] https://www.thriftyfun.com/tf80568815.tip.html.
[15] http://japanlifeandreligion.com/2008/03/21/the-joy-of-frugality/.
[16] https://www.dayalbagh.org.in/radhasoami-faith/basic-concepts.htm.

Simplistic Model of Bhutan—A Happy Little Kingdom

Known as a '*Happy Little Kingdom*', Bhutan may be considered as one of the world's least-developed countries. However, the country is determined to protect its unique culture and safeguard its social values by entrenching them in the ways that the wider world may understand and respect in quantifiable measures. By developing measures of progress that account for the country's social, cultural, and environmental assets as well as its economic development, the country is following through on the 1972 declaration made by His Majesty King Jigme Singye Wangchuck: 'Gross National Happiness (GNH) is more important than Gross National Product'. Article 9 of the constitution of Bhutan lays special emphasis on happiness and well-being of its citizens through GNH. It is explained as a 'multi-dimensional development approach seeking to achieve a harmonious balance between material well-being and the spiritual, emotional and cultural needs of society' with four pillars of GNH like promotion of equitable and sustainable socio-economic development, preservation, and promotion of cultural values, conservation of the natural environment, and establishment of good governance.[17] The very basis of ensuring the above is their frugality based behavioural practices, leaving little scope for consumerism or uncalled for development which may spoil the very fabric of their carefully crafted traditions, culture and natural habitat. '*We have to think of human well-being in broader terms*', said Lyonpo Jigmi Thinley, Bhutan's home minister and ex-prime minister. '*Material well-being is only one component. That doesn't ensure that one is at peace with one's environment and in harmony with each other*'.[18]

Many of the policy experts and economists may dismiss this very concept as '*naïve idealism*' or a '*compulsive frugal behavioural choice*' due to limited resources, as it against the trend visible in the study conducted at Harvard which indicates that '*relative wealth becomes more important than the quality of life*'. Similarly, Diener and Diener (1995), in their study entitled 'The Wealth of Nations-Income and Quality of Life' compared 101 countries on 32 indicators of quality of life and universal human values (e.g. happiness, social order, and social justice). The study concluded that wealth significantly correlates with better quality of life.

---

[17] https://www.gnhcentrebhutan.org/what-is-gnh/history-of-gnh/.
[18] http://sepiamutiny.com/blog/2005/10/04/bhutans_gross_n/.

However, *The New York Times* (4 October 2004) of United States reported the statement of a senior Bhutanese official that 'The goal of life should not be limited to production, consumption, more production and more consumption',[19] as the connection between '*how much we have*' and '*how good we feel*' is not very definite and is contextual in nature. Money, power, and prestige can be the symbols of success but not necessarily the symbols of a healthy environment, peace, and happiness. A popular saying states that they can provide food but not appetite, clothes but not beauty, a house but not a happy home, books but not wisdom, medicines but not health, beds but not sleep, partners but not love and above all the luxuries of life but not happiness in life. Therefore, Bhutan believes in being in close touch with environment and appraising it positively as the path for global well-being.

In spite of the above analysis, it is well acknowledged that limited resources, alcoholism, and the pressures of modernization were mounting up day by day in Bhutan, particularly among the young generation and in the name of frugality, the developmental needs of the country cannot be disregarded. But still, the case of Happy little kingdom of Bhutan has lead the consumer—centric western world to rethink on the core culture of such countries with a strong spiritual base and minimalistic lifestyle and they are still trying to find out the hidden force behind positive environmental appraisal and subjective well-being in them.

Similarly, another eastern religion, Taoism believes that nature is self-regulating and self-expressing. It transforms in a consistent manner as per need. Therefore, one has to be receptive to nature with minimal interference in natural forms. All things with their transformations and changes are considered to be self-regulating, self-expressing in their natural form. Hence, frugal way of life in harmony with nature can allow the spontaneity of nature in varied forms.

Dayalbagh—An Indian Hermitage for Frugality

If one wants to look at the environmentally healthy and sustainable, economically self-reliant, politically untouched, socially and psychologically satisfying and spiritually escalating temple of wisdom, worldliness,

---

[19] https://www.nytimes.com/2005/10/04/science/a-new-measure-of-wellbeingfrom-a-happy-little-kingdom.html.

and spiritual vision, one has to know the case of *Dayalbagh*, which literally means '*Garden of the Merciful*'. Located within the important urban centre, Agra, the city of Tajmahal, in India, this small place has been in existence since 1915 as a self-contained township of the followers of *Radhasoami* sect. It also emphasizes that humans should make it the aim of their life to develop their faculties of all the three kinds—physical, mental, and spiritual, throw off all lethargy and develop '*work is worship*' as a mission of life.[20] To procreate this mission, the '*Garden of Merciful*' was tastefully created and crafted following principles of frugality, to provide the inhabitants, an honest simple living, employment through small scale industries, high-level educational facilities, health facilities, and a simple and sustainable residential complex.

Dayalbagh has been described as an eco-village as far as peace and quietude is concerned. It is also seen as an ideal city as far as providence of basic amenities and cleanliness is concerned. Taken together, it is a perfect live example of frugality in action. Active efforts are being made for environmental conservation and thereby addressing the likely adversities of changing climate which may include initiatives in 'harnessing the renewable energy through Solar thermal and Solar Photovoltaic (SPV) power plants and using Solar Thermal Cooking systems powered by 7 Distributed Roof-Top Solar PV power Plants aggregating to a total of 1000 kW'.[21] The value-based holistic education system imparted in Dayalbagh University targets to reach the '*last, the least, the lowest and the lost*' through Community-driven participatory development, social service based village adoption programme, frugal solutions to meet their developmental needs and creation of Rural Economic Zone while connecting them to national and international markets.

The sixth revered leader of Radhasoami faith, His holiness Mehtaji Maharaj pointed out that neither does wealth flow here nor people live with any deprivation; neither are there any big palaces and mansions, nor any dilapidated huts either; neither is anyone great or big nor anyone small and insignificant, and if anyone is honoured here more than others,

---

[20] https://www.dayalbagh.org.in/.
[21] https://unescochaircbrsr.org/pdf/presentation/Story_ofDayalbagh_Educational_Institute_Anand_Mohan.pdf.

it is he who works better than others. Dayalbagh belongs to every resident, while no resident has any kind of property here.[22] On hearing this and looking at the surroundings and people, Jawaharlal Nehru, the first prime minister of India commented '…countries advance not an account of their size but by the manner its people lead their lives and by the character they possess and by the skill of their hands and intelligence they possess to do things. If India would make progress, it would do so because such people live here and not because thirty-six billion people live here'.

Simple and frugal lifestyle with limited usage of consumer luxuries is the motto here so that physical, mental, and spiritual faculties can be developed in harmony as well as sustainability of environment can also be maintained. People are encouraged to participate in physical exercise followed by social service in agricultural fields to remain physically healthy and productive. *'Fatherhood of God and brotherhood of mankind'* is the basic principle of this faith which teaches universal brotherhood pervading all prejudices and superstitions regarding caste, creed, nationality, or colour. Frugality is practiced even for occasions like birth, death, or marriage with minimal rituals and expenditures. Happiness, where material well-being is concerned, is mostly a question of values and the world would be spared of many a quarrel and consequent misery, if only a right value system is inculcated. Besides cultivation of spiritual values, the social and environmental values are also given due importance. It is believed that frugal lifestyle alone leads to harmony between human and human, between human and natural environment and can make the world a better place to live in.

The case of *Dayalbagh* signifies balancing of the economic and environmental indicators and shows the path of sustainable lifestyle reflected through '*better worldliness*'. At one side, there is a well-developed motorized fast transportation system as practiced in developed and many of the developing countries, to meet the movement requirements, at another side, there is a behavioural choice of *Dayalbagh*, to allow very limited motorized transportation within the campus in spite of the premises housing about Five thousand followers of *Radhasoami* sect, thus avoiding the possible air and noise pollution. Bicycles and *e-rickshaws* are the major mode of transportation over there. Besides, there is a major emphasis on frugality in electricity consumption too as the very process of electricity

---

[22] https://www.dayalbagh.org.in/publications/publication_files/Publication.html.

generation is not friendly to environment. The most interesting feature is the good affording capacity of most of the people living over there and it is their behavioural choice to enjoy the limited worldliness within a sacred and spiritually guided behavioural setting. Limited enjoyment of consumer products, to the extent that person may not became a slave of it, is the beauty of that socio-behavioural setting which promotes positive environmental appraisal (Corral-Verdugo et al., 2011; Rishi, 2009).[23]

The whole world is now realizing the need for using bicycles to conserve oil and avoid air and noise pollution, besides fitness reasons. The whole world is now realizing the ills of electronic and social media and its associated social, psychological, and health effects on the generation going to take the charge of the world in the time to come. At the same time, the whole world now is also realizing the need for social support network and strong family bonds to obtain solace, sympathy, and support for emotional release which counters the psychological distress arising out of complexities of life, rising temptations, and unrealistic desires. Not only this, the whole world is also increasingly realizing the need for spiritual wisdom to obtain contentment and satisfaction with the present and move towards psychological well-being and total quality of life from different perspectives. Frugality-based lifestyle is the answer to proceed from materialism to supra-materialism and ultimately achieve the balance between the two for preserving the global climate (Rishi, 2009).[24]

However, mere usage of consumer goods to serve the necessity is not materialism. A complete dependence of people on consumer products and psychic attachment with the comfort associated with their usage, supplemented by using them as a status symbol, without any consideration for its long-term impact on the environment, is what materialism is all about. Everyone agrees that there was a time of simplicity, with fewer consumer products and things, fewer technologies and a slower pace of life. This simplicity now seems to be a blessing in disguise as it has some invisible virtues. Verhelst (1990) describes them as 'vitality, a taste for life, human qualities and a sense of the sacred that leave a lasting impression of hope'. He further describes that harsh social struggles do not get in the way

---

[23] Rishi (2009). Environmental Issues Behavioural Insights. Jaipur, Rawat Publications.
[24] Rishi (2009). Environmental Issues Behavioural Insights. Jaipur, Rawat Publications.

of 'a sense of celebration, human tenderness and ultimately, the desire for social harmony'. The case of Dayalbagh also embodies the voluntary simplicity/frugality model for ultimate satisfaction.

BEYOND MATERIALISM—A FRUGALITY MODEL

'It is great wealth to a soul to live frugally with a contended mind'[25] remarks Lucretius. Materialism is a never-ending process of consumptive behaviour, which commonly leads to physical dependence and psychic attachments, besides being ecologically unsustainable in a long run. Research conducted in both developed and the developing world proves that materialism has always been a driving force to ecological crisis and there have been consistent efforts to slow down its pace in order to realize the vision for a sustainable future (Norgaard, 1995). Further, materialism promotes an attitude of careless and irresponsible exploitation of natural resources with a mere utilitarian approach towards natural environment. The time has come when we should give a serious thought to '*post materialism*' thought processes for the cause of environment and well-being (Taylor, 2017) and transform our lifestyles. In the similar context, concepts like '*reflexive modernity*' given by German sociologist Ulrich Beck also became popular explaining materialism '*in the realms of nature, society and personality*' otherwise, it may lead to '*invisible and irreversible global risks*' (Beck et al., 2003). Subsequently, the concept of '*sustainable materialism*' came up which talks about an integration of conservational practices in the '*material flows*' (Schlosberg, 2019) of consumption and production practices for the cause of sustainability of resources.

The fundamental premise behind all the above is an attitude of frugality and '*minimalism*', which may also be called as '*Supra-Materialism*'.[26] It is a frugality-based behavioural choice wherein, in spite of the viability of resources to get them, they are being used sparingly as and when required with minimal mental attachment to them.

'*Bhagwad Gita*', the sacred book of India, also narrates the importance of supra-materialism for eternal happiness. It states that one should always concentrate on action and not on the fruit of that action to remain focused on his inner self without disturbing his equilibrium through

---

[25] https://www.brainyquote.com/authors/lucretius-quotes.
[26] Rishi (2009). Environmental Issues Behavioural Insights. Jaipur, Rawat Publications.

success or failure and pleasure or pain. This state of equilibrium is also known as '*Yoga*' or the state of restraint from worldly desires and control over mind and body. Yoga is basically a skill involved in performing on action. This is one of the stages of supra-materialism in which in spite of performing all worldly action using materialistic things, a person is above them as far as attachment to them is concerned. Control over senses stabilizes the mind. Where this desire remains unsatisfied, it results in anger which may cloud one's thinking abilities, impair the memory and destroy the logic or reassuring power, resulting in loss of mental stability and equilibrium. On the other hand, desire and anger are insatiable and are the biggest enemies which are all consuming and polluting that enwrap the world and overpower the wisdom. The *Bhagwad Gita* also states how reduction of desires can be a key to happiness for humans. If the strength of wisdom is compared, it flows from the lowest level, i.e. from sense organs to mind, a little higher level to intellect, a little higher level to spirit, the most powerful superior and mightier of all which can subjugate the mind with its might and crush the desire with its supreme power and make possible the communion with god almighty. The frugality-laden behavioural practices have the potential to create conducive environment for physical, social, psychological as well as spiritual development of humankind.

There is a need to move from materialism to supra-materialism, rationally balancing the needs for consumer goods, for achieving the ecological and social sustainability. Sheer materialism can promote global warming, environmental degradation, and natural resource depletion on one hand and possessiveness, dependence, and psychic attachment on the other, which may not provide peace and contentment with the life. On the other hand, total restraint from all materialistic possessions like Buddhist monks and many others may also be practically difficult for people with limited religious inclinations and prove to be detrimental to growth and development of third world where developmental indices are low. Therefore, restraining from materialism cannot be stated as the only source of happiness and well-being in the third world. One should rather move from materialism to supra-materialism which can be described as '*Better Worldliness*' (a high-quality worldly life of contentment, humanitarian values and compassion and an attitude of frugality to use available material resources with unity of purpose, thought, and action).

In order to connect supra-materialism to sustainability and to minimize Climate Change impact through alternative lifestyle, '*Society for*

Preservation of Healthy Environment and Ecology and Heritage of Agra (SPHEEHA)' was established in Agra city of India in the year 2007. Its prime objective was '*to work for the management and implementation of sustainable and holistic solutions for the pressing environment and ecological needs*' of the city.[27] Likewise, there are several other notable organizations in India for social and ecological change such as *Fiinovation* (a Delhi-based CSR consultant in the social development sector with an emphasis on social activities and sustainability), Institute for Sustainable Communities (ISC), SoulAce (a research and advisory firm operating in the CSR & Development Sector space in the South Asia region working with Corporate, NGOs, Government and Funding agencies), PARFI Foundation, *Avani* (a community built on the principles of sustainability and local empowerment in the *Kumaon* region of the Indian Himalayan state of Uttarakhand) just to name a few.[28]

Frugality-Based Innovations for Sustainability

'Businesses must move away from the top-down organizational hierarchies….and transform themselves into social enterprises built on bottom-up, agile models on collaboration. Jugaad Innovations can enable your entire employees, customers, and partners—to make significant contributions and drive hyper-growth', remarks Marc Benioff.[29]

We can well appreciate that most of the product design and innovation takes place keeping the high-end consumers in mind, making them unaffordable and unsuitable for the larger population groups of the developing world. The Concept of frugality is meaningless if it is not able to provide creative and low-cost solutions to serve the needs of modern-day society in an equitable, sustainable, and environmental friendly manner. Frugal innovations are the answer to that, intending to transform the global society through low-cost need-based solutions and eventually raising their standard of living. Basu et al. (2013) describe Frugal Innovation as a process which prioritizes the necessities and societal milieu of inhabitants of the under-privileged class at the forefront while developing '*appropriate, adaptable, affordable, and accessible services and products for*

---

[27] https://spheeha.org/.
[28] https://thecsrjournal.in/csr-5-organizations-india-social-environmental/.
[29] https://fdocuments.in/document/praise-for-jugaad-innovation.html.

emerging markets'. Authors further explain that most of the developed countries were unable to hold the frugality-based behavioural practices in the past, leading to resource crunch, ecological destruction, and a plethora of other undesirable consequences. von Janda et al. (2020) presents frugal innovation as having four dimensions: '*cost of consumption, sustainability, simplicity, and basic quality*'. Broadly speaking, some of the desirable features of frugal innovations are given in the following Fig. 4.2.

One of the most sought after features of frugal innovation is its inclusiveness, i.e. to benefit the larger section of society and not just the few privileged ones with more economic resources (Prahalad & Mashelkar, 2010). It stands similar to Gandhian model which is driven towards '*more from less for many*' and shifting the focus from '*low price, low performance*' to '*low price, high performance*' (Altenburg, 2009) products and services through limited/indigenous resources and customizable as per local needs. Young Indian population with limited affordability and high aspirations presents a fertile ground for companies to try out their creative

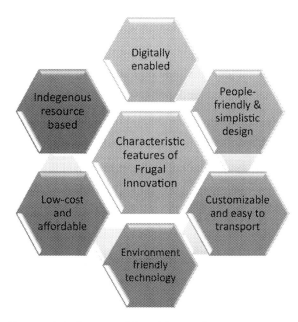

Fig. 4.2 Characteristic features of frugal innovation

but frugal ideas on the ground with not much compromise with the quality of the product (Tiwari, 2017) which can be called as '*affordable excellence*' (Tiwari, 2017).

Hence, the time has come when we have to think beyond the thin creamy layer of the society, as needs of the underprivileged class, which is proportionately much larger in number, are equally important to be addressed. On this premise, the term 'Jugaad' in Hindi language was coined to offer, frugal, flexible, and fast track solutions to local needs with limited resources in an innovative and inexpensive manner.[30] It is also known as '*creative improvisation*'. India has pioneered in some of the very popular frugal products, such as basic level 'TATA Swach' water purifier, light weight solar lantern and most useful of all is 'Jaipur Leg' which is a rubber-based prosthesis to help physically challenged people. Dayalbagh Educational Institute, in Agra city of India, is also exploiting such '*Jugaad*'-based frugal innovation ideas of students and faculty into practice while using cost-effective, in-house design and technology. Facilitation of digital education, telemedicine, e-commerce networks in tribal areas of central part of India through '*last mile mobile connectivity*' are some of the examples of such 'Jugaad Innovation' which has empowered rural/tribal women to establish their '*nano-enterprises through frugal innovation and low-cost input*'.[31]

[30] https://www.dei.ac.in/dei/files/notices/2020/JUGAADNEW.pdf.
[31] https://www.dei.ac.in/dei/files/notices/2020/JUGAADNEW.pdf.

Some other examples are 'earth-friendly power grid, rooftop solar panels and biomass plants to generate electricity' to make the university campus self-contained as far as energy needs are concerned. Re-creation of old and heavy-on fuel vehicles into solar power based self-recharging electric vehicles is another example of Frugal innovation through '*Jugaad*' as reflected in the image showing rooftop solar panel.[32]

Dayalbagh Educational Institute is working in partnership with '*The Prakash Lab*' under '*Stanford Woods*'—Institute of Environment, at Stanford University, USA under the guidance of bioengineer and now called frugal scientist, Prof. Manu Prakash, who is considered as a pioneer in Jugaad innovations. His foldable paper origami based microscope ('*foldscope*') is a very popular equipment providing scientific exposure to remotely located rural children. Foldscope has the capacity to magnify 2000 times using inexpensive lenses, which is sufficient enough for much sought after health care solutions in Malaria-affected countries with limited infrastructure and also number of educational applications.[33] Besides, some of his other frugal innovations were intended to create low-cost health care solutions for the management of COVID-19 and some of them are in the form of '*Do it Yourself*' (DIY) like 'a converter to make full-face snorkel masks into reusable personal protective equipment, or PPE; guidelines for decontaminating N95 masks; a universal remote for controlling ventilators from outside a patient's room; a ventilator built with abundantly available parts; a simple filtration test setup for face masks' and few others.[34] Another frugal-sustainability start-up of northern part of India, which is first of its kind is of '*Kagzi-bottles*' (bottles made of paper) which is 100% compostable and trying to create a sustainable solution to plastic menace.[35]

Many of the multinational companies are also looking forward to embrace '*Jugaad*'-based frugal innovations to serve the resource-constrained population. Ananthram and Chan (2019), conducted a study on Indian multinational companies to explore the institution framework

---

[32] https://www.dei.ac.in/dei/files/notices/2020/JUGAADNEW.pdf.

[33] https://125.stanford.edu/frugal-science/.

[34] https://engineering.stanford.edu/magazine/article/lab-takes-frugal-science-approach-covid-19.

[35] https://www.thebetterindia.com/256160/kagzi-bottles-samiksha-ganeriwal-noida-startup-compostable-sustainable-paper-packaging-made-in-india-plastic-pollution-innovation-c24/.

and enablers for creation and sustenance of '*Jugaad*'—a creative and indigenous frugal innovation. A complex interplay of institutional factors and organizational characteristics operated in the process which forced the multinationals to innovate frugally leading to '*Jugaad*' outcomes. Authors confirmed that individuals who engage in pro-ecological and frugal actions are also likely to practice altruistic and equitable behaviours. Thus, a person that practices sustainable behaviour, not only engages in one kind of actions but tends to act in an integrated pro-environmentally manner. Thus, frugal innovation has the capacity to realize '*social sustainability*' and subsequently move the world towards achievement of Sustainable Development Goals (Khan, 2016).

Psycho-Social Correlates of Frugality

Frugality research is cross-cutting across psychology, sociology as well as marketing, due to its interconnected nature and applications across individuals and societies (Goldsmith & Flynn, 2015). There are limited number of studies exploring behavioural and psychological dimensions concerning frugality. The existing studies have primarily assessed attitude towards frugality, frugality in relation to diversity, altruism, pro-environmental behaviour, sustainability behaviour, intrinsic motivation, personality factors and so on and so forth. Cross-cultural variations among such behavioural patterns are also prominent. In a study, relationship between frugality attitude, environmental concern, and intention to engage in energy consumption/consumer behaviour was analysed among Japanese sample population. Attitude towards frugality was found to be positively correlated with reduced electricity and gas consumption. Results were interpreted in line with the Japanese concept '*mottainai*', meaning '*respect for resources*' which was presumed to be closely associated with attitude towards frugality (Fujii, 2006). Hence, educating people for frugality has the potential to promote sustainability and pro-environmental action. However, empirical research is required to substantiate the same.

In American context, a study by Goldsmith and Flynn (2015) explored motivating factors behind behavioural practice of frugality. The trait of frugality showed significant positive correlations with self-control and independence while negative correlations with materialism. In a study on respondents from New Zealand by Todd and Lawson (2003), it was found that lower levels of frugality were associated with individualistic tendencies and value systems while higher levels of frugality were associated with value systems associated with collectivism. However, they finally concluded that frugality, which is associated with restraint and conservation of resources, is more of a life style than a value system.

Another study in Mexico, by Corral-Verdugo et al. (2016), found a significant relationship between frugality and equity-linked sustainable behaviour and intrinsic motives of satisfaction, autonomy, and self-efficacy. Results established that frugality and associated sustainability behaviour/pro-social/pro-environmental behaviour is intrinsically motivated and self-determined. In some behavioural studies, frugality and sustainability behaviours have been found to be related with happiness (Corral-Verdugo et al., 2011). By showing an association between pro-environmental and frugal behavioural choices and preferences for altruism and equity, Tapia-Fonllem et al. (2013) reflected that individuals generally have an integrated and holistic approach towards environment. Roczen et al. (2010) highlights the importance of nature appreciation in promoting frugality. Satisfying and engaging experiences with natural environment can ignite mental energy, promote physical health and act as buffer against stressful situations (Kaiser, 2013).[36]

Based on the analysis of a complex array of psychological and behavioural traits in relation to frugality, sustainability, and pro-environmental behaviour across countries, it is required that frugality should be recognized as one of the personality traits. Younger generations, going to take charge of planet earth in the time to come, should be encouraged, trained, and appreciated in frugality in both developing as well as developed countries irrespective of their affordability, for the cause of sustainability.

Conclusion

Frugality is a way of life. It has the potential to redefine our life priorities and helps us to move towards a more inclusive and sustainable society. It also establishes harmony between living systems of planet earth, which acts as the pre-condition for sustainable development. However, there is a need for transdisciplinary perspective to appreciate and apply this concept for the betterment of lives of a sizable population across the globe. The

---

[36] https://cedmcenter.org/wp-content/uploads/2013/04/Rebound-Think-Piece-Kaiser.pdf.

synergy between countries, disciplines, and professions can help embrace frugal innovation in a more meaningful manner as an enabling 'driver of progress in achieving sustainable solutions' (Basu et al., 2013).

A popular phrase in India in *Sanskrit* language '*Vasudhaiv Kutumbakam*' aptly describes the concept of frugality for sustainability and leads to '*better worldliness*'. It means, treat planet earth as your extended family and show your compassion to every living being, including animals and plants along with human beings. Human Beings should justify the 'being' by rising above the 'doing' and 'having' trap of modern-day materialism (Parikh and Gokarn, 1991). This is what inclusivity of frugal innovation means with the intention to promote better world order worldwide and a happier and healthier planet as well as people. To attain this state, there is a need for redefining our needs and moving from '*lifestyle of abundance*' to '*lifestyle of sufficiency*' which is possible through introspection and asking few questions to oneself like, '*How much is enough for me?*' and '*how much the planet can support?*' Through this approach, we can move from '*high consumption-huge impact lifestyle*' to '*modest consumption-sustainable lifestyle*' as highlighted by Durning (1991).

Synthesizing the eastern and western perspectives of the world by linking the earning potential and living potential may be desirable for the creation of inclusive world, where compassion, contentment, and co-creation through frugal innovation for the larger good of people and planet is being practiced. Complimenting and supplementing the social, cultural, and environmental values across countries will enrich and develop the world as a whole which is one of the acceptable agendas under CSR funding, to take it further towards Sustainable Development Goals.

References

Alexander, S. (2011). *The voluntary simplicity movement: Reimagining the good life beyond consumer culture*. Available at SSRN 1970056.

Altenburg, T. (2009). *Building inclusive innovation systems in developing countries: Challenges for IS research*. Edward Elgar Publishing.

Ananthram, S., & Chan, C. (2019). Institutions and frugal innovation: The case of Jugaad. *Asia Pacific Journal of Management*. https://doi.org/10.1007/s10490-019-09700-1

Basu, R. R., Banerjee, P. M., & Sweeny, E. G. (2013). Frugal innovation. *Journal of Management for Global Sustainability*, 1(2), 63–82.

Beck, U., Bonss, W., & Lau, C. (2003). The theory of reflexive modernization: Problematic, hypotheses and research programme. *Theory, Culture & amp; Society, 20*(2), 1–33.
Bouckaert, L., Opdebeeck, H., & Zsolnai, L. (2008). *Frugality: Rebalancing material and spiritual values in economic life* (Vol. 4). Peter Lang.
Bove, L. L., Nagpal, A., & Dorsett, A. D. S. (2009). Exploring the determinants of the frugal shopper. *Journal of Retailing and Consumer Services, 16*(4), 291–297. https://doi.org/10.1016/j.jretconser.2009.02.004
Chhokar, J. S. (2007). India: Diversity and complexity in action. In *Culture and leadership across the world* (pp. 1005–1054). Psychology Press.
Corral-Verdugo, V., Frías-Armenta, M., & García-Cadena, C. H. (2013). *Introduction to the psychological dimensions of sustainability.*
Corral-Verdugo, V., González-Lomelí, D., Rascón-Cruz, M., & Corral-Frías, V. O. (2016). Intrinsic motives of autonomy, self-efficacy, and satisfaction associated with two instances of sustainable behavior: Frugality and equity. *Psychology, 7*(5), 662–671.
Corral-Verdugo, V., Mireles-Acosta, J., Tapia-Fonllem, C., & Fraijo-Sing, B. (2011). Happiness as a correlate of sustainable behavior: A study of pro-ecological, frugal, equitable and altruistic actions that promote subjective wellbeing. *Human Ecology Review, 18*, 95–104.
De Young, R. (1996). Some psychological aspects of reduced consumption behavior: The role of intrinsic satisfaction and competence motivation. *Environment and Behavior, 28*(3), 358–409.
Diener, E. D., & Diener, C. (1995). The wealth of nations revisited: Income and quality of life. *Social Indicators Research, 36*(3), 275–286. https://doi.org/10.1007/BF01078817
Durning, A. (1991). Limiting consumption. *The Futurist, 25*(4), 10.
Fujii, S. (2006). Environmental concern, attitude toward frugality, and ease of behavior as determinants of pro-environmental behavior intentions. *Journal of Environmental Psychology, 26*(4), 262–268. https://doi.org/10.1016/j.jenvp.2006.09.003
Goldsmith, R. E., & Flynn, L. R. (2015). The etiology of frugal spending: A partial replication and extension. *Comprehensive Psychology, 4*, 9–20.
Jackson, T. (2012). The challenge of sustainable lifestyles. In *State of the world 2008* (pp. 73–88). Routledge.
Kahneman, D. (2011). *Thinking, fast and slow.* Macmillan.
Kaplan, R. (1991). Environmental description. In *Environment, cognition, and action: An integrated approach* (p. 19). Oxford University Press.
Khan, R. (2016). How frugal innovation promotes social sustainability. *Sustainability, 8*(10), 1034.

Lastovicka, J. L., Bettencourt, L. A., Hughner, R. S., & Kuntze, R. J. (1999). Lifestyle of the tight and frugal: Theory and measurement. *Journal of Consumer Research, 26*(1), 85–98. https://doi.org/10.1086/209552

Nathawat, S. (1975). Yoga and behaviour therapy. *Perspectives in Yoga* (pp. 63–93).

Norgaard, R. B. (1995). Beyond materialism: A coevolutionary reinterpretation of the environmental crisis. *Review of Social Economy, 53*(4), 475–492.

Parikh, J., & Gokarn, S. (1991). Consumption patterns: The driving force of environmental stress.

Parkins, W., & Craig, G. (2011). Slow living and the temporalities of sustainable consumption. In *Ethical consumption: A critical introduction* (p. 189). Routledge.

Prahalad, C. K., & Mashelkar, R. A. (2010). Innovation's holy grail. *Harvard Business Review, 88*(7–8), 132–141.

Radjou, N., & Prabhu, J. (2015). Frugal innovation: How to do more with less. *The Economist*.

Roczen, N., Kaiser, F. G., & Bogner, F. X. (2010). Leverage for sustainable change: Motivational sources behind ecological behavior. In *Psychological approaches to sustainability: Current trends in theory, research and applications* (pp. 109–123). Nova Science.

Roiland, D. (2016). Frugality, a positive principle to promote sustainable development. *Journal of Agricultural and Environmental Ethics, 29*(4), 571–585.

Rosca, E., Arnold, M., & Bendul, J. C. (2017). Business models for sustainable innovation—An empirical analysis of frugal products and services. *Innovative Products and Services for Sustainable Societal Development, 162*, S133–S145. https://doi.org/10.1016/j.jclepro.2016.02.050

Sandler, R. L. (2009). *Character and environment: A virtue-oriented approach to environmental ethics*. Columbia University Press.

Schlosberg, D. (2019). From postmaterialism to sustainable materialism: The environmental politics of practice-based movements. *Environmental Politics*.

Shoham, A., Gavish, Y., & Akron, S. (2017). Hoarding and frugality tendencies and their impact on consumer behaviors. *Journal of International Consumer Marketing, 29*(4), 208–222.

Tapia-Fonllem, C., Corral-Verdugo, V., Fraijo-Sing, B., & Durón-Ramos, M. F. (2013). Assessing sustainable behavior and its correlates: A measure of pro-ecological, frugal, altruistic and equitable actions. *Sustainability, 5*(2), 711–723.

Taylor, S. (2017). Moving beyond materialism: Can transpersonal psychology contribute to cultural transformation. *International Journal of Transpersonal Studies, 36*(2), 147–159.

Tiwari, R. (2017). Frugality in Indian context: What makes India a lead market for affordable excellence? In C. Herstatt & R. Tiwari (Eds.), *Lead market*

India (pp. 37–61). Springer International Publishing. https://doi.org/10.1007/978-3-319-46392-6_3

Todd, S., & Lawson, R. (2003). Towards an understanding of frugal consumers. *Australasian Marketing Journal (AMJ), 11*(3), 8–18.

Verhelst, T. G. (1990). *No life without roots: Culture and development.* Zed Books.

von Janda, S., Kuester, S., Schuhmacher, M. C., & Shainesh, G. (2020). What frugal products are and why they matter: A cross-national multi-method study. *Journal of Cleaner Production, 246,* 118977. https://doi.org/10.1016/j.jclepro.2019.118977

Westacott, E. (2016). *The wisdom of frugality.* Princeton University Press.

Zwarthoed, D. (2015). Creating frugal citizens: The liberal egalitarian case for teaching frugality. *Theory and Research in Education, 13*(3), 286–307.

CHAPTER 5

Integrating CSR with Climate Change and Sustainability

INTRODUCTION

Corporate organizations across the globe are an integral part of the society. Right from the inception, business houses have been partnering beneficially with the communities as they have contributed to generate employment by creating products and services along with generating profits and working for organizational growth. However, their endeavours are not supposed to be limited to economic progression as they are expected to be partners in developmental processes by acting in a socially responsible manner towards inclusive and sustainable growth. Therefore, they are supposed to play a significant role by not only providing much needed monetary resources but also through their expertise in the management of large social and environmental projects in an efficient way, and thus contributing to the developmental needs of the society. Hence, there has been a continuously increasing pressure on the corporate bodies to contribute meaningfully towards the issues of multiple stakeholders, society at large, and the natural environment. Progressive companies have already proven that they can differentiate their brands, images as well as their products and services if they take responsibility for the well-being of the societies and environments within which they operate. Those companies are leveraging Corporate Social Responsibility (CSR) in a manner

© The Author(s), under exclusive license to Springer Nature Singapore Pte Ltd. 2022
P. Rishi, *Managing Climate Change and Sustainability through Behavioural Transformation*, Sustainable Development Goals Series, https://doi.org/10.1007/978-981-16-8519-4_5

that generates substantial returns to their businesses and provides sustainable competitive advantage (Porter & Kramer, 2006). The benefits are manifold regarding improved sales, improvement of customer loyalty and brand perception, and the ability to attract and retain employees. By leveraging these benefits, the organizations can improve their bottom line and top line, attract more capital, and unleash better economic value.

Though the whole concept of social responsibility has generic voluntary connotation, the negligent behaviour of many of the corporates towards society and environment, coupled with large-scale corporate governance frauds, to earn profits in an unlawful manner, was observed in many countries. At the same time, globalization, financial crises, unethical practices, misuse of corporate power, and environmental disasters have forced organizations to consider the consequences of their actions on environment, local community, and the broader society (D'Aprile & Mannarini, 2012). Even the persistent interest and scrutiny by non-governmental organizations, media, and governments have led to a demand for organizations to take responsibility for their actions and address the societal problems (Calabrese et al., 2013). It has led us to rethink the need for a regulatory framework to govern CSR provisions in a systematized manner so that the role of corporate organizations in societal development and environmental conservation is more clearly defined.

This chapter takes a deeper look into CSR and SDGs' integration in a manner that it can help address environmental issues and climate change adaptation as well as promote achieving SDGs through integration of socio-psychological concerns at micro- and macro-level.

Levels of CSR—A Behavioural Analysis

Unprecedented challenges in economic, environmental, and social fronts require emergence of sustainable development as a unifying principle for addressing these issues through contribution of CSR (Macagno, 2013). Further, Palazzo and Richter (2005) have differentiated between three levels of CSR: instrumental, transactional, and transformational as reflected in Fig. 5.1.

The instrumental level refers to the skills and competencies that are required to deliver quality products and services to its consumers/customers. It is the most basic level of social responsibility and considered as equivalent to hygiene factors of Herzberg's two factor

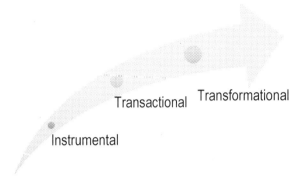

Fig. 5.1 Levels of CSR

theory of motivation (Herzberg, 1964).In simpler terms, it means that providing quality products and services to the society is one of the foremost social responsibility of any corporate organization, without which, it cannot claim any further stake in the arena of CSR in broader sense of the term. It also corresponds to Kohlberg's pre-conventional stage of moral development (Kohlberg, 1985) in which people adopt morality because they don't want to attract any punishment for not being moral and in this context, corporate organizations do not want to attract any legal action for producing low-quality products.

Further to that is the transactional level, wherein the organizations must comply with the legal and moral rules of their societal context. In the current context of India, it is desirable that they should run their CSR activities as per Section 135 and Schedule VII of Indian Companies Act (2014)[1] as they don't want to forego their image of responsible organization in the eyes of consumers/society.[2] It is like Kohlberg's conventional stage of moral development, where morality is what significant others around you are expecting from you. If other competitive organizations are performing their CSR as per regulatory framework/norms, you are also

---

[1] http://ebook.mca.gov.in/Actpagedisplay.aspx?PAGENAME=17923.

[2] Ministry Of Corporate Affairs—FAQ on CSR cell. (n.d.). Retrieved September 29, 2020, from http://www.mca.gov.in/MinistryV2/faq+on+csr+cell.html.

supposed to follow the same as you want to remain recognized as responsible organization as per motivators of Herzberg's Two Factor Theory (Herzberg, 1964).

The third and the desired, transformational level is where the organizations project their willingness to transcend self-interest for the common good and support the societal well-being in a proactive manner with full commitment. It matches with Kohlberg's principled stage of moral development and Maslow's concept of self-actualization (Maslow, 1943). It is this level of CSR that Indian government wishes to achieve through CSR law in which innovation, sustainability, scalability, and impact assessment are the integral parts of CSR projects. Its primary objective is that the money spent in CSR projects can be meaningfully utilized for the need-based sustainable development of the society, taking the steps forward towards achieving of the SDGs by 2030.

Mirvis (2011) says that there is a progressive development of companies with increasing levels of sophistication in order to reach the sustainability and CSR goals which may range from an elementary to engaged, innovative, and integrated level and the highest level is creative/innovative edge. It can be a game changing approach towards CSR and sustainability—transpiring out of incessant interface between a company and its environment that inspire organizational learning. With every new stage, the engagement with societal concerns gets increasingly more open and its connections with stakeholders get more collaborating and reciprocal. As a result, the organizational structures and practices/processes to accomplish corporate responsibilities become more erudite and synchronous with routine business operations (Haropoulou, 2013).

CSR—A Stakeholders' Analysis

In the twenty-first century, the roles of corporate organizations and society are going through most profound shifts. In the last three decades, it is bringing a major change in the purpose of doing business that has a substantial impact on individuals, communities, society, and the planet at large and may positively affect the lives of the future generations too. The companies, whose major objective had been to maximize profits for the benefit of their shareholders, have shifted focus on serving the interest of stakeholders, natural environment, and the society at large.

The DNA of CSR 2.0 model by Visser (2010) combines aspects of both stakeholder theory and sustainable development theory through the concepts of Connectedness (C), Scalability (S), Responsiveness (R), Duality (2), and Circularity (0). In the model, Connectedness explains relations with multi-stakeholders. Scalability explains that CSR programmes must be carried out by a company on a vast scale and for a long-term duration. Responsiveness explains that a company must be responsive to deliver the needs of the community. Duality explains that a company is not only responsible towards economic aspect but also towards environmental and social issues and Circularity explains that the company must conform to sustainability. The author further states that his model can be considered as 'interconnected, with non-hierarchical levels, representing economic, human, social and environmental systems, oriented towards sustainability/responsibility: economic sustainability and financial responsibility; human sustainability and employee responsibility; social sustainability and community responsibility; and environmental sustainability and moral responsibility (Visser, 2010)'.

Benyaminova et al. (2019) says that CSR should be staunchly stakeholder-driven and proclaim that companies should adapt their operations according to their dependency on the resources that their significant stakeholders possess. This, therefore, clearly establishes the stake of a business organization in the physical, social, and environmental well-being of a society, of which they are a part, to enhance their quality of life and well-being. Therefore, many companies are paying genuine attention to the principles of social/environmental responsiveness. As per the UN Global Compact—Accenture CEO Study (2010),[3] CSR has been declared as an 'important' or 'significant' factor for organizations' future success, as perceived by 93% of the 766 participating CEOs from all over the world.[4] The profit motive behind business hardly helps attain sustainability in the long run. Thus the organizations have started shifting focus on all the relevant stakeholders, instead of the shareholders alone.

McQueen (2019) highlights the crucial role of social activism in shaping people's attitudes, policy intervention, and subsequently influencing legislation, in the context of the western world. He further argues

---

[3] https://d306pr3pise04h.cloudfront.net/docs/news_events%2F8.1%2FUNGC_Accenture_CEO_Study_2010.pdf.

[4] https://dash.harvard.edu/bitstream/handle/1/9887635/cheng,ioannou,serafeimCorporate%20Social%20Responsibility%20and%20Access%20to%20Finance.pdf?sequence=1.

that shifting from fossil fuels to greener fuels to avoid the disastrous impacts of climate change can also be addressed through such activist movements as it seems to be the responsible response to an otherwise unsuccessful effort of business and government both. Role of children and youth activists can be furthermore effective in this regard for long-term headway towards a sustainable and socially responsible society. Examples include Greta Thunberg[5] and the sisters Fatima and Amna Sultan.[6]

Box 5.1: The Young Ambassadors of Climate Change
Greta Thunberg from Sweden started campaigning when she was aged 15, after winning a climate change essay contest. Further, she protested in front of the Swedish parliament to force them to meet the carbon emission targets of the 2015 Paris Agreement. Further, she could mobilize more than 20,000 students across the world to join Friday missing lessons campaign and addressed the UN climate conference in September 2019. She was also named Time magazine's person of the year.

Fatima and Amna Sultan, the young sisters were on a mission to inspire and empower people for climate change and sustainability by using art as the mode of bringing people together for this cause. They represented their gender and generation to ensure a sustainable future. They have an immense experience of travelling around the globe and establishing libraries for marginalized communities. They delivered key notes in around 40 international conferences and popularized their book '*Everything is possible*'.

Fatima and Amna intend to inspire people through art to spread the message that everyone should accomplish their dreams with full determination and age/time is no bar to that. It eventually led them to present their business on the TV show Dragons Den.

Corporates must rethink their institutional arrangements and devise ways of applying the principle of what the European Union calls 'subsidiarity' in redesigning the system. This applies equally within business and within society. If corporates can devise ways of empowering communities in collaboration with NGOs and then attempt their linking with CSR execution, especially for enhancing quality of social and environmental life at grass root level, it can be their sustainable contribution to

---

[5] https://www.bbc.com/news/world-europe-49918719.
[6] https://twosistersonamission.com/.

5 INTEGRATING CSR WITH CLIMATE CHANGE AND SUSTAINABILITY

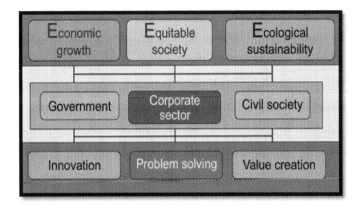

Fig. 5.2 CSR for sustainability through stakeholders' integration (Based on Wipro's Sustainability report, 2010–11)

society at large. The same idea is reflected by Azim Premji in his sustainability report of Wipro[7]: *The new world order needs three Es—economic growth, equitable society, and ecological sustainability. We also believe that business must play a crucial role in making this happen. Government and civil society are equal stakeholders in this mission, but as a crucible of innovation, problem-solving and value creation, the business sector is uniquely positioned to make a vital difference* (Fig. 5.2).

CSR principles require organizations to expand their commitments beyond profit generation and obeying the minimum required by the law to increase the social and environmental sustainability of the community in which they operate. After the behavioural and stakeholders' analysis of CSR, its integration with SDGs is one of the robust steps to proceed with the sustainability agenda. Many governments and not-for-profit sectors recognized that working towards achieving SDGs can help to increase profits besides taking the sustainability agenda ahead. CSR, with principle of sustainability ingrained in it, is strongly instrumental in achieving the SDGs in a novel way, across the globe.

---

[7] https://www.wipro.com/content/dam/nexus/en/sustainability/archieves/wipro-sustainability-report-2010-11.pdf.

CSR and Sustainable Development Goals

In September 2015, all 193 Member States of the United Nations embraced a plan to end hunger, poverty, inequality, and injustice by 2030, to protect our planet. This goal is ensuring a better future for all and how the world would be in the next 15 years. The multi-year plan of the UN Global Compact was to drive business awareness and actions in support of achieving 17 Sustainable Development Goal (SDGs) by 2030 for people, planet, and towards a better future. These goals chiefly focus upon channelizing resources to combat poverty and create a life of dignity and generate opportunity for all (United Nations, 2018).[8] SDGs are aimed to achieve 17 strategic global goals, out of these 9 goals— No poverty, Zero hunger, Life below water, Life on land, Peace, Justice and strong institutions, Good health and well-being, Clean water and sanitation, Decent work and economic growth, and Reduced inequalities are related to various qualifying activities under schedule VII of Indian company law, which can be undertaken by corporate organizations under CSR.These goals adopt a rational approach for sustainable growth at all levels with the concept of inclusive development of society and environment. The goals provide direction by which maximum can be obtained for community well-being with minimum resource utilization. The sustainability goals impart flexibility to both the countries and the organizations alike to develop plans to accomplish the set milestones.

Sustainable Development Goals (SDGs) serve as an allusion framework in improving CSR by contributing to sustainable development (Schönherr et al., 2017). India too, with other countries, has signed the declaration on the 2030 agenda for sustainable development and thus adheres to the 17 SDGs and the 169 targets. India has further linked SDGs with its National Development Goals and has set ambitious goals for implementation of these SDGs. India till 2030, will develop the policies and the decision-making at the national level addressing the SDGs with self-planning and huge investments that are must for inclusive growth. Both, the guidelines by the department of public enterprises of India and the CSR law have channelized finances towards CSR to attain the goals of sustainable development.[9] However, it is possible only if organizations

---

[8] https://www.un.org/en/sections/issues-depth/poverty/.

[9] https://assets.kpmg.com/content/dam/kpmg/in/pdf/2017/12/SDG_New_Final_Web.pdf.

methodically manage and measure the effects of corporate on sustainable development as a pre-condition to demonstrate the contribution to the SDGs.

From Sustainable Development to Sustainability Behaviour

The term 'sustainable development' was popularized in 1987, by the World Commission on Economic Development (WCED) in the famous report, Our Common Future (Diamond, 1996) in which it was explained as 'the development that meets the needs of the present without compromising the ability of future generations to meet their own needs'. Further, sustainable development necessitated the concurrent adoption of environmental, economic, and equity principles (3Es), challenging the deeply ingrained perception that environmental/climate integrity and social justice is in dispute with the economic affluence. (Martinuzzi & Krumay, 2013) have stated that social responsibility is an obligation of an organization to contribute to the broader goals of the community and thus have directly associated the CSR concept with sustainable development. As a result, strategies for corporate sustainable development integrate their ecological integrity through corporate environmental management, social equity through CSR, and economic prosperity through value creation (Elkington, 1998; WCED, 1987).

Ecological environments are susceptible to deplete due to the population, development pressures as well as their limited generative capacity. As per the Stern Review on the Economics of Climate Change (Stern et al., 2010), ecological sustainability could become the central social responsibility challenge for business. Overall, organizations are encouraged to continually integrate environmental and social approaches into their core business activities (Carroll & Shabana, 2010). Hence, environmental integrity of corporates expects them not to erode the land, air, and water resources during their production process and at the same time, work for their conservation under their CSR/corporate budget for biodiversity conservation, afforestation, efficient waste management practices, eco-friendly production process, and environmentally safe products which is possible through a well laid environmental management policy. In this manner, they can contribute for corporate sustainable development from an ecological perspective. In India, Schedule VII of Section 135 of Companies Law 2014 is having a specific focus on ensuring environmental sustainability encompassing all the above as a part of one of the mandatory CSR provisions to be executed within 2% of their CSR budget.

Sustainable Development operates at the macro-level, but it is not possible until and unless the micro-level sustainable behaviour is promoted at the organization as well as at the individual front. Small individual actions towards sustainability can take us towards larger goal of corporate sustainable development and further to that working towards it at the global front. Bansal (2005) states that application of principles of environmental integrity, social equity and economic prosperity to products, policies, and practices is a necessary condition towards sustainable development. Behavioural processes adopted by organizations are also equally important to contribute to micro-level sustainability behaviour. A comprehensive model integrating these principles of sustainable development with corporate functionality is given in Fig. 5.3.

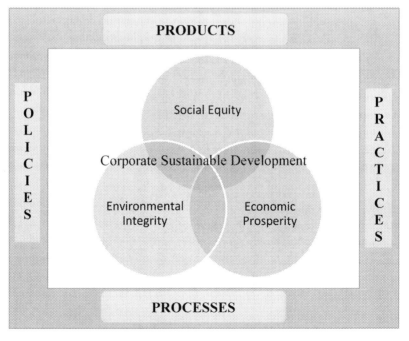

Fig. 5.3 Framework of corporate sustainable development

The SDGs associated with environmental integrity are clean energy, clean water and sanitation, climate action, life below water, and life on land. Broadly, corporates are expected to execute these dimensions through inclusion in strategic corporate policy as well as CSR policy as their implementation within the corporate production process as well as outside the domain of in-house environmental management, under the provisions of applicable CSR policy/law is desirable, to ensure the integrity of environmental dimensions in a sustainable manner (Table 5.1).

Sustainability includes not only the sustenance of ecological environment but also the social environment. Corporate organizations are crafted out of existing social systems of the country/nation where they function with the objective of serving the needs of people (consumers) and hence, accountable to the society. Hence, corporate sustainability necessities ensuring social equity which means providing equal access to resources and opportunities to people and at the same time, supplementing the responsibility of national governance structure by extending a helping hand in basic needs' satisfaction. SDG 1 and 2 talk about poverty alleviation and ensuring availability of nutritious food while SDG 3 and 4 are intended to ensure quality of life through health and hygiene and quality education. The universal meaning of sustainability includes not

Table 5.1 Corporate sustainable development and associated SDGs

Dimension of CSD	Associated SDG
Environmental integrity	GOAL 6-Clean Energy
	GOAL 7-Clean water and sanitation
	GOAL 13-Climate Action
	GOAL 14-Life below water
	GOAL 15-Life on land
Social equity	GOAL 1: No Poverty
	GOAL 2: Zero Hunger
	GOAL 3: Good Health and Well-being
	GOAL 4: Quality Education
	GOAL 5: Gender Equality
	GOAL 10: Reduced Inequality
	GOAL 16: Peace and Justice, Strong Institutions
Economic prosperity	GOAL 8: Decent Work and Economic Growth
	GOAL 9: Industry, Innovation and Infrastructure
	GOAL 12-Responsible consumption and Production
	GOAL 17: Partnerships to achieve the Goal

only the present but future generations too. Social equity, thus, encompasses within itself, the needs of indigenous people, marginalized classes, castes, and communities too across generations through SDGs 5 and 10, in which focus on reducing gender and other inequalities is made within the broader framework of SDG 16, for peace and justice as well as meaningful creation and functioning of strong institutional systems to ensure all the above (Table 5.1).

In India, the disparities in population are extremely high on almost all fronts. Hence, Schedule VII of Companies Law 2014, especially meant for implementation of mandatory CSR in eligible companies, point (*i*) eradicating extreme hunger and poverty; (*ii*) promotion of education; (*iii*) promoting gender equality and empowering women; (*iv*) reducing child mortality and improving maternal health; (*v*) combating human immunodeficiency virus, acquired immune deficiency syndrome, malaria, and other diseases; (*vii*) employment enhancing vocational skills; and (*viii*) social business projects, have been especially kept to ensure social equity in which corporates are supposed to play a very significant role through their mandatory CSR activities as per their CSR policy framework.

The dimension of economic prosperity is envisioned to uphold a reasonable productive capacity of organizations and individuals and thus promoting the quality of life in their existing social system (Holliday et al., 2002). It is possible only by creation of value through production and marketing of high-quality goods and services to help people elevate their standard of living (Bowman & Ambrosini, 2000). Value creation is further enhanced through product innovation and enhancement of cost efficiency and providing cheap product (Conner, 1991). Global markets within the system of open economy and healthy competitive work culture, encouraging innovation, efficiency, and wealth creation are central features of sustainable economic prosperity. It is inherently tied to the dimensions of social equity and environmental integrity (Schmidheiny & Timberlake, 1992) and WCED (1987). Therefore, as per principles of corporate sustainability management, business value creation and enhancement are possible only if dimensions of environment and social equity are integrated with it.

Indian CSR section in company's law encompasses all the SDGs as well as business sustainability principles with mandatory legal framework, to make it happen on the ground and encourages its execution by generating healthy completion among corporates through National CSR

Awards under the ambit of Ministry of Corporate Affairs, Government of India.[10] The awards are intended to recognize the initiatives of corporates for inclusive growth and sustainable development, especially those which have positively impacted both business and society by adopting a strategic approach to CSR through participatory CSR programme involving different stakeholders. Further, they also aim to identify the companies working for transformation of business operations by integrating sustainability as part of their core business model and implement procedures for conservation and sustainable management of the biodiversity and ecosystem in the value chain. Further, identifying innovative approaches and employing application and technologies is also positively evaluated to build robust CSR programmes to advance the foundation of inclusive and sustainable development. Hence, awards are constituted to broadly cover all the SDGs by recognizing the corporate initiatives to further human development, economic development, social welfare, and environment and sustainable development through well designed and impactful CSR projects and programmes.

The Strategic Management, CSR and Corporate Sustainability Analysis presents a systematic framework that provides a holistic view of the importance of sustainability aspects to a business and allows them to be incorporated at various management levels. The significance of sustainability concerns at different levels of governance, as well as opportunities and challenges linked to sustainable development, can be established at normative, strategic, and operational management level (Baumgartner, 2014).

SOCIAL PROCESS REENGINEERING (SPR) FOR RESPONSIBLE BUSINESS PRACTICES

With the companies' law 2014 of India, CSR is gradually becoming the heart and soul of organizations. To some, it seems like another tax to meet the mandatory requirements of doing business while to others, it's a strategic opportunity to have an edge over others by doing business in a manner that is sustainable and responsible, surpassing the limits of success. The key point which makes a difference between these two categories of business professionals is the way they are able to connect with the society

---

[10] http://www.mca.gov.in/Ministry/pdf/NationalCSRAwards_11052017.pdf.

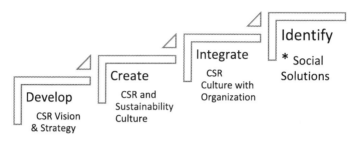

Fig. 5.4 Social process reengineering in CSR

and understand/practise their social responsibility. It is the social system consisting of people, the basic units of society, which has given them the opportunity to serve them. Without having the humanized connection with people, the end users of all carefully crafted products and services by the business world, no business can survive. Here, Social Process Reengineering (SPR) comes into picture (Frank, 2010),[11] which is, in short, the understanding and application of social software in the context of CSR, in order to increase the efficacy of CSR practices for the larger benefit of stakeholders. Figure 5.4 explains the process undergoing SPR starting with the development of CSR vision and strategy to creating the CSR and sustainability culture to integrating the culture with organizational culture and identifying the social solutions for process reengineering.

Hence, Process Reengineering, the buzz word of the twenty-first century is closely linked to society and its fundamental unit, the people. Now the question arises, who are these people? The customers currently having the high-end spending capacity to buy the products and services of the business world or much beyond that? What wrong has been done by people who have remained unable to taste the fruit of development since generations? Do we not have the responsibility to address their needs, as per our capability? Can the government alone be held responsible for taking care of developmental needs of society like India? Shouldn't we all have our own individual roles to play, to sail the boat of development in a country like India where developmental needs are immense. To answer these series of unanswered questions, the opportunity has now come when each and every business can work on their

---

[11] http://tractionsoftware.com/traction/permalink/Blog1316.

share of the social responsibility under section 135 of Indian company's law 2014 pertaining to CSR through systematized project-based approach and taking up qualifying CSR activities as per Schedule VII under CSR funds.

Challenges for CSR and Need for SPR

The biggest challenge which comes in the way of social responsibility is that business world is dominated by technologists as professionals. On the other hand, they are also addressing the needs of social sector, in which a very different skill set is required to plan and execute social projects. Besides, being receptive to genuine social needs at the ground level they are also expected to empathize with the real challenges which people are facing while addressing their harsh ground realities. It further substantiates the need for identifying social solutions to process reengineering in the business world. Figure 5.5 gives the overview of how SPR can be executed in CSR context.

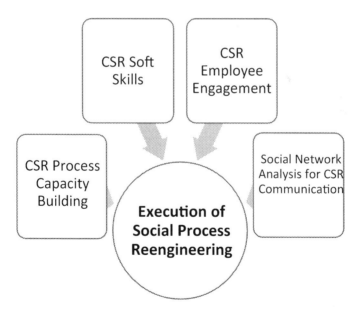

Fig. 5.5 Execution of SPR in CSR context

CSR Process Capacity Building

For executing SPR in the context of CSR, the first and foremost step is to build the capacity of professionals undertaking CSR activities, so that project identification, planning, and execution is effectively done and it is able to give competitive advantage to companies as well. The CSR professionals are supposed to have the following capacities:

- to understand the importance of CSR in business context to identify innovative, sustainable, and scalable CSR projects which are need-based and can give strategic advantage to companies within the ambit of CSR clause.
- to develop a realistic time plan and execute CSR projects as per government guidelines.
- to identify credible implementing agencies having expertise in the field of project in a timely manner.
- to periodically monitor and evaluate the implementation process and ensure its timely completion.
- to get its third party impact assessment done from professional experts of the field.

There is a need for companies to understand the importance of capacity building for effective execution of CSR so that the benefits can reach at the right place in an impactful and sustainable manner. A lot of capacity building programmes by professional organizations are being organized to strengthen the knowledge base of professionals handling CSR-related works. Besides that, there is a need for another skill set which plays a more crucial role in ensuring success in the field of CSR execution and its efficacy and that is CSR soft skills as described below.

CSR Soft Skills

Soft skills are in general perceived as the skills, which do not appear to be at the surface as hard technical skills but have the strong potential to influence the tip of the iceberg while remaining at the latent level like a strong bottom of the iceberg. In a similar manner, CSR soft skills include a set of skills which are required for the professionals dealing with the diverse CSR projects, in order to have a better connect with the CSR team at the organizational level and also with the local functionaries at the ground level where CSR projects are being executed, in a manner that the efficacy of CSR is enhanced. This is possible only when the professionals have the ability to exhibit effective and context appropriate CSR leadership which can be reflected through the ability:

- to patiently listen to the local issues and convert them into implementable high impact projects
- to engage employees in a manner that CSR activities are not considered as alien to routine business operations but remain the integral part of it.
- to work in a team surpassing bureaucratic organizational structure to plan, execute, and evaluate the impact of CSR projects.
- to build social sustainability culture
- to have a strong ground connect with the beneficiaries and stakeholders

Dedicated capacity building programmes in CSR soft skills for corporate CSR team can help enhance employee engagement and contribute towards impactful and sustainable CSR activities.

Employee Engagement and CSR
CSR is linked to employee engagement through reduced costs due to increased employee retention as well as improved reputation in the eyes of employees besides many other reasons. CSR programmes impact the drivers of employee engagement, e.g. employee behaviour and motivation, stakeholder attitudes and behaviours, e.g. potential employees, and positively influence business outcomes, e.g. employee productivity and retention (Knox et al., 2005). Companies are realizing that CSR can bring competitive differentiation and favourable positioning in the war to attract best talents. Developing and implementing a CSR strategy by engaging employees is a unique opportunity but in practice, a limited percentage of businesses engage their employees on the company's CSR objectives and initiatives and thus a noteworthy opportunity is lost.

Employee engagement in CSR initiatives can be a powerful recruitment and retention tool in an environment where the war for talent is shaking up whole industries. Tower Perrin Global Workforce Study

(2007–2008)[12] found that CSR is the third most important driver of employee engagement overall, and that an organization's reputation for social responsibility was an important driver for both engagement and retention among all age groups except 18–24 years of age. An important opportunity rests with CSR's potential to influence employee engagement and, consequently, the positive business outcomes that go along with an engaged workforce.

Thus, there is a pressure on organizations to increase CSR activity in order to recruit and retain a top quality workforce because employees see a socially responsible organization as a fair organization and reciprocate this fairness through dedication, loyalty and increased productivity.

The above analysis proves the ability of CSR to serve as an effective tool to maximize companies' return on investment in social responsibility initiatives through employee engagement:

- by using an input–output approach to formulate and implement CSR initiatives, and subsequently evaluate and manage CSR outcomes.
- by understanding and fulfilling employee needs related to CSR, targeting strategic employee segments.[13]
- by enabling employees to be co-creators of CSR value (Bhattacharya et al., 2007).

Corporate social responsibility has made significant growth over the last decade in terms of policy creation, organizational and management guidelines, key performance indicators, best practices, and creation of core values. Now employees want to feel good about their employment choice by working in a responsible organization. Rupp et al. (2013) mentioned that employees make distinct judgments about their organization's CSR efforts. These perceptions provide evidence regarding the fulfilment of psychological needs and acts of social responsibility or irresponsibility on the part of the organization, which can trickle down to affect employees' subsequent attitudes and behaviours.

---

[12] https://c.ymcdn.com/sites/www.simnet.org/resource/group/066D79D1-E2A8-4AB5-B621-60E58640FF7B/leadership_workshop_2010/towers_perrin_global_workforce.pdf.

[13] https://www.researchgate.net/publication/241561787_CORPORATE_SOCIAL_RESPONSIBILITY_AS_AN_INTERNAL_MARKETING_STRATEGY.

CSR initiatives should be integrated and internalized by the companies in such a manner that they are placed at the centre of the business and not merely as an add-on to it. Corporations that have understood this new *mantra* towards corporate success have internalized the benefits flowing from responsible corporate citizenship and can hope to reap rich dividends in terms of improved corporate reputation, gaining consumers' trust, employee motivation and retention, and favourable market positioning, and so on (Dey & Sircar, 2012). Engagement of employees in CSR activities of the company will surely generate interest and pride in CSR work and promote transformation from CSR as an obligation to socially responsible corporate in all respects of their functioning.

Social Network Analysis (SNA) for CSR Communication
Communication is an integral part of CSR. The very basis of social Networks is communication, verbal or written or virtual. Potential benefits of Social Network Analysis (SNA) in making CSR communication effective, is a pertinent component of Social Process Reengineering. However, it must be remembered that communication is merely a tool to acquire strategic as well as a competitive advantage from being socially and environmentally responsible. If a company over focuses on the instrumental aspect of CSR, and ignores the importance of its transactional and transformational perspective (Palazzo & Richter, 2005), CSR practices may be misperceived by stakeholders as just one of the company's marketing strategies. Subsequently, it might prove to be counterproductive to the very objective of project and programme-based CSR activities. Hence, CSR Communication may act as a '*double-edged sword*' which might bring certain benefits for the socially and environmentally responsible company, but at the same time, may also be perceived as an activity motivated by instrumental and transactional CSR in order to achieve economic gains for the company which is not desirable as per requirements of mandatory CSR under section 135 of Indian Companies Law.

Every organization is connected with its stakeholders through various social networks. CSR professionals publish information as per their perception of its importance which they want to highlight. They presume that others will also perceive it similarly which might be a misleading assumption on their part. Stakeholders are fully aware of the fact that organizations try to communicate only their good deeds. Hence, more important is to understand the complexities of social network ties with their internal and external stakeholders and the importance and benefits of managing those ties (Grzesiuk, 2017).

Regarding CSR communication, it is essential for the organization to have a well-dedicated team which continuously updates the stakeholders about ongoing and planned CSR activities and seeks their inputs to enrich the same. In this manner, employee CSR connection is maintained and appropriate opportunities to engage employees and other stakeholders in CSR pursuits can also be realized. Since social networks are extensively used by stakeholders, conducting a social network usability survey of employees, and accordingly channelizing the information, can virtually organize stakeholders and motivate them to contribute for CSR activities in different capacities.

Taken together, SPR broadly encompasses within itself the fourfold components namely CSR process capacity building, to build up the knowledge base and acquaint the concerned people of the company with existing social groups and process in order to plan and execute CSR at the cognitive level; CSR soft skills, in order to equip them with required social, behavioural, and managerial skills for effective execution at conative level; CSR employee engagement, to establish the employee–CSR connection at affective level and utilize that for better selection and retention of employees and finally, Social Network Analysis (SNA) for effective CSR communication by utilizing social networks to connect the stakeholders on a common platform while sharing the CSR-related information and seeking their inputs for enhancing efficacy and promoting better connect.

Integrating CSR with Environment and Climate Change

It is well known that global natural environment has been blemished in the process of industrialization and uncalled-for human intervention. Hence, CSR can be instrumental to compensate this loss in the manner, which government, on its own, may not be able to practise in an efficient manner. One major norm for any action/programme to qualify for CSR, especially in Indian CSR law is that it must be unique and innovative and disqualifying element will be that it should not be the replication of an already ongoing government programme. CSR has widened the arena of environmental sustainability by providing funds in natural resource management, climate change, sustainable practices like solar energy equipment, integrated watershed management, rivers and lake conservation, pollution abatement programmes, emission reduction, waste management, organic agriculture, development of biodiversity parks, and many more. Environmentally sustainable activities for corporates include green products, green production processes, eco-labelling, eco-branding, eco-marketing, awareness generation in employees and communities about environment, reducing plastic use and power consumption, CDM/Carbon Trading, reducing GHG emissions, and climate adaptation.

Thus, it is expected from the industries that they act in a socially responsible manner along with acting sustainably to meet the expectations of their relevant stakeholders and also promote inclusive growth (Oginni & Omojowo, 2015). However, CSR practices promoting sustainability seem to be unclear regarding morals and principles guiding business operations. Thus, there exists a need to determine the focus of CSR practices of present corporate organizations regarding sustainable development goals for a safe and inclusive growth. Renowned international actors at the UN Conference on Sustainable Development, Rio + 2, and many others, called upon the organizations to direct policy and resources towards sustainable development goals to streamline economic activities posing a threat to the environment and fail to reduce poverty level across the world.[14] With times, there has been a shift observed in the use of terms, wherein sustainability appears to be more encompassing and thus seems to be gaining favour over social responsibility. For implementing CSR in environmental sustainability arena, there is a need for highly skilled natural resource/environmental management consultants, having expertise in conducting base line surveys/need assessment for CSR projects keeping in mind the prevailing ecological conditions. They can also contribute to conducting research to explore the potential for the development of environmental sustainability programmes, preparation of project action plans, their implementation, and finally conducting the impact assessment of CSR activities in the environmental and natural resource management arena.[15]

---

[14] https://sustainabledevelopment.un.org/content/documents/1030rwanda.pdf.

[15] A guidance document for CSR in natural resource and environment management sectors, Kinhal, Rishi and Pandey, 2015. Inhouse IIFM publication.

CSR has occupied a position in the present organizations' strategic framework by playing a crucial role in addressing the issues emerging out of climate change as well as achieving the SDGs in a novel way. Climate change causes range of developmental deficits across the globe which can be addressed through robust and systematically planned CSR practices. Anthropogenic factors substantially influence the natural ecosystem like endangering biodiversity, depleting ozone layer, accrual of greenhouse gases, concentration of waste, deforestation, and increase in environmental toxicity (Doering, 2002). With reference to developing countries, Puppim de Oliveira and Jabbour (2017) state that environmental improvements can be attained through legal enforcement, supply chain pressure, and voluntary commitment within CSR funding from business organizations. According to Hishan et al. (2020) climate change, being a long persisting issue of global concern affecting corporate organizations, civil societies, and leaders, needs a global solution. It is believed that since businesses run in society, societal permissions must be equated to legal recognitions for resource utilization, be it social or natural or human. It further suggests the application of social contract theory to manage the industry-civil society interface to resolve social and environmental problems (Ray, 2013), in a study conducted in India states the need for institutional frameworks to monitor the impact and efficacy of CSR projects. It is possible only if employees are involved in CSR execution at all the levels and have a reasonably good awareness and indulgence in CSR and sustainable development.

As Indian CSR projects are mostly operating in rural areas, ground level understanding of issues underpinning rural dynamics of the country in the context of sustainability is very crucial. A need for knowledge management system in CSR and sustainability to learn from the best practices as well as failures can help maintain a balance between the struggle for growth and inclusive development. Mandatory CSR guidelines for Indian organizations have the potential to attain sustainable development, if implemented and executed in letter and spirits.

CSR IN THE ERA OF COVID-19—PSYCHO-SOCIAL CONCERNS

The review of comprehensive literature highlights the fact that the next key factor which could potentially shape the fate of CSR is the novel coronavirus—COVID-19. It has not only caused worldwide severe economic crisis in which over 350 million people lost their jobs, but also opened the doors for extending CSR beyond triple bottom line through integration of psycho-social dimensions in sustainable development process. Resultantly, the role of corporate organizations is going to be transformed as the socio-economic catalysts contributing for the holistic development of the society, taking human behavioural factors into consideration. Some of the key activities under CSR as linked to COVID-19 are elaborated below:

CSR for Pushing Digital Interface and Remote Learning

It is very important to know how COVID-19 has impacted the steps forward towards achievement of SDGs in 2020 and beyond. United Nations Department of Economic and Social Affairs, in its economic analysis policy brief-78 regarding achieving the SDGs through the COVID-19 response and recovery states, that there is a likely drop in global GDP in the range of 3.2–5.2%, which is the largest one since the Great Depression, and even worse than the 2008–2009 global financial crisis. There is a likelihood that around 35–60 million could have been pushed into poverty, and the efforts of last many years across the globe could have been retrogressive in declining this global trend.[16] International Labour Organization, in its recent report says that there has been a serious threat to the livelihood of around 1.6 billion people working in the informal sector and many of them do not even have access to social protection.[17] Besides, 10 million children across the world could face severe nutritional deficiencies along with acute food insecurity which is likely to double growing to 265 million.[18] More than 90% of world's students, i.e. 1.6 billion children and youth, were affected by suspension of education due to lockdowns.[19] A large percentage of them were unable to access e-learning due to limited/no connectivity, resulting in primary school

---

[16] https://www.un.org/development/desa/dpad/publication/un-desa-policy-brief-78-achieving-the-sdgs-through-the-covid-19-response-and-recovery/.

[17] https://www.ilo.org/global/about-the-ilo/newsroom/news/WCMS_743036/lang--en/index.htm.

[18] https://www.wfp.org/news/covid-19-will-double-number-people-facing-food-crises-unless-swift-action-taken.

[19] https://www.un.org/development/desa/dspd/wp-content/uploads/sites/22/2020/08/sg_policy_brief_covid-19_and_education_august_2020.pdf.

dropout rates equivalent to mid-1980s, causing worry due to life-long learning deficits, and educational inequalities across generations.[20] The above scenario opens the door for large-scale educational interventions concerning digital learning under CSR, to restart the steps towards SDG 4 and 10.

An array of standard operating procedures (SOP) was adopted in corporate organizations to contain the outbreak of pandemic. Some of these interventions have been counterproductive for the corporate organizations, as shutdown of companies, restrictions on non-essential businesses and corporate travel, social distancing, staying at home directives, mandatory temperature scanning, and quarantine of high-risk and sick employees and additional health and safety concerns involved huge investment on their part. Besides, total lockdown of cities, restrictions on social gatherings, continuing information overload on virus prevention measures, and many other directives under disaster prevention act have left many of the ongoing CSR activities and people in utter dismay and the need for integrating health and wellness as one of the prime components of CSR is gaining utmost importance across all parts of the world.

Integrating Health, Wellness, and Social Responsibility

It is expected that the effect of COVID-19 on CSR is likely to be unprecedented making the call for a radical reform of businesses' health and safety policies, more urgent than ever. Most studies have shown that not every health pandemic outbreak is a matter of global CSR concern across organizations. The much more important factor contributing to a pandemic or global health crisis is that it must have the capacity to produce global socio-economic interruption of a ghastly proportion (Auld et al., 2008) as it has happened in the case of COVID-19. The pandemic must have been up to the task of mobilizing self-defendants and at the global arena, it must have a gripping value that reaches the political arena of worldwide humanitarian groups and multinational corporations (Robinson et al., 2012). Hence, companies have a moral responsibility to promote systemic changes through collaborative health vulnerability assessments by engaging with national/state level health and safety agencies (Antwi, 2020).

---

[20] https://www.un.org/development/desa/dpad/publication/un-desa-policy-brief-78-achieving-the-sdgs-through-the-covid-19-response-and-recovery/.

The crisis in general populace and global health has led to and continues to encourage changes in a manner that justifies the preferences of social responsibility advocates. The enormous impact of COVID-19 on business enterprises and the global economy has sparked an unparalleled and inconceivable transition in the CSR framework and processes as the world continues to struggle to restrict the virus (Livingston et al., 2020).

While contributing positively and emotionally to the reduction of the socio-economic burden of COVID-19, there is a moral responsibility of corporate organizations to reject radical free-market philosophies that take precedence over the cost of employee health and well-being. Hence, there is a need for drastic reforms of health and safety policies of businesses as it is their legal duty to reform existing risk assessments in cooperation with state agencies for health and safety investments with world-class checks.

COVID-19 and the CSR Transition

In India, the corporate sector has assumed the responsibility of providing food assistance to some of the worst affected people who lost their daily livelihood due to lockdowns and it was in addition to humanitarian assistance, in-kind and cash donations, transportation facilities, food supplies, etc., to other needy members of society (Gentilini et al., 2020). Some of the big CSR concerns and challenges under COVID-19 are salary changes, layoffs, redundancies, absenteeism, and salaries for sick and stay-at-home personnel (Buera et al., 2020). Companies will also deal with the disinfection of companies or industries; among others, changing operating hours, splitting open spaces (Williamson et al., 2020).

The common apprehension is that pressures of survival and limited finances may force companies to look for short-term gains, resorting to even scams and transgression, resulting in compromised CSR spending. However, to the contrary, the observation is that many companies across the world have repelled unscrupulous corporate practices during pandemic and indulged proactively in responsive CSR activities, extending support to the needy people. But still, the scenario varies across companies in both east and west, with some, exploiting the opportunities with mindful and insightful approach to CSR, while others compromising with mandatory CSR activities due to financial constraints/using pandemic as an excuse for not investing in CSR (Buera et al., 2020).[21] There are many cases in India, the United Kingdom, and in many other countries where transformation of business operations to respond to immediate needs even by compromising profitability was observed. This responsive initiative led to addressing the everyday increasing needs for ventilators, personal protective equipment, hand sanitizer, etc., in many nations, with some of them even adopting charitable modes to fulfil the need for these products. Vodafone, a telecommunications company gave free unlimited data access to financially vulnerable customers.[22] Another support is being extended by supermarkets in the United Kingdom by having specified working hours for the senior citizens along with operating food banks with the contributory support from voluntary groups like Fair share which work for fighting hunger and tackling food waste.[23] UK tea brand PG tips also were uniting with other social organizations to address social seclusion and isolation among elderly people during pandemic linked lockdown (Jones, 2020).[24] Waiving interest on overdrafts by banks during pandemic phase was also practised.[25]

The Growing Optimism

So, looking at the above observations, there are reasons to be optimistic that Covid-19 pandemic is likely to accelerate post-pandemic CSR in the long run. The companies understand their business sustainability is reliant on accomplishing a balance between financial prosperity and establishing congruence with multiple stakeholders. So, the question is not whether to invest under CSR budget or not but how to invest in CSR to draw strategic advantage out of it along with achieving mutually advantageous and symbiotic sustainability objectives. So, it is a lesson of togetherness, to be learned, raising people's expectancy regarding social responsibility of business, with resilient obligation for CSR and well laid strategies

---

[21] https://doi.org/10.1016/j.jbusres.2020.05.030.
[22] https://www.bbc.com/news/technology-52066048.
[23] https://fareshare.org.uk/what-we-do/.
[24] https://metro.co.uk/2020/04/07/.
[25] https://www.ncbi.nlm.nih.gov/pmc/articles/PMC7241379/.

for efficient implementations. CSR walk is going to take over CSR talk under closely scrutinized informed consumers and public, for corporate sustainability.

Conclusion

Looking at the fast-emerging challenges in the business world, society, and the environment at large, there has been an imperative prerequisite for the corporate world in almost all the countries to wilfully connect themselves with sustainable development goals following the principles of corporate responsibility. SDGs are formulated to have a positive impact on the biosphere, society, and economy. Implementation of SDGs by the corporate world can help them achieve the triple bottom line—social, environmental, and financial.

Pandemic has changed the way we see our world and has given a colossal opportunity to corporates to strategically place and align their business operations and CSR activities in a manner that they could draw strategic advantage out of it along with making genuine proactive as well as reactive contribution to society. This is a golden opportunity for corporates to shift towards authentic and important concerns of the society and make a positive change. It may either be in the form of alternative energy or fossil fuel reduction or waste management strategies or promotion of digital education infrastructure or strengthening health support network in remote areas as per need or collaborating with voluntary organizations at grass root level for running nutritional support programmes and food banks for people who have lost their livelihoods in the wake of the pandemic.

Active employee engagement as well as stakeholders' participation is crucial for impactful and effective CSR projects. Corporate organizations are expected to partner with academic institutions with strong knowledge base, to assist them in managing their CSR projects in a professional manner starting with need assessment to connecting them with professional implementing agencies to impact assessment. They can also help the organizations in capacity building and providing them expert professionals for undertaking CSR activities. There is a need to equitably share the global responsibility and create a multi-stakeholder and multicentre responsible milieu. This will help develop a sense of responsibility as well as contribution among all stakeholders to take the sustainability agenda ahead.

Corporates are not obliged to their shareholders only but are answerable to the society as well. There have been concerns over number of social responsibilities taken up by corporates in different countries. Current pandemic has made us realize that there is a need for equitable CSR across the world. There is also a need for awareness to underline the fact that CSR-related sustainability concepts are still way behind known to managers of companies and public institutions. Important steps for cognizance must be included while improving the mindset of people in the organizations. The need to transform and integrate CSR into corporate sustainability behaviour can be another goal for organizations integrating micro- as well as macro-level behavioural transformation efforts for sustainability and pro-climate action.

References

Antwi, H. A. (2020). *Beyond COVID-19 pandemic: A systematic review of the role of global health in the evolution and practice of corporate social responsibility* [Preprint]. In Review. https://doi.org/10.21203/rs.3.rs-40212/v1

Auld, G., Bernstein, S., & Cashore, B. (2008). The new corporate social responsibility. *Annual Review of Environment and Resources, 33*, 413–435.

Bansal, P. (2005). Evolving sustainably: A longitudinal study of corporate sustainable development. *Strategic Management Journal, 26*(3), 197–218. https://doi.org/10.1002/smj.441

Baumgartner, R. J. (2014). Managing corporate sustainability and CSR: A conceptual framework combining values, strategies and instruments contributing to sustainable development: Managing corporate sustainability and CSR. *Corporate Social Responsibility and Environmental Management, 21*(5), 258–271. https://doi.org/10.1002/csr.1336

Benyaminova, A., Mathews, M., Langley, P., & Rieple, A. (2019). The impact of changes in stakeholder salience on corporate social responsibility activities in Russian energy firms: A contribution to the divergence/convergence debate. *Corporate Social Responsibility and Environmental Management, 26*(6), 1222–1234.

Bhattacharya, C. B., Sen, S., & Korschun, D. (2007). Corporate social responsibility as an internal marketing strategy. *Sloan Management Review, 49*(1), 1–29.

Bowman, C., & Ambrosini, V. (2000). Value creation versus value capture: Towards a coherent definition of value in strategy. *British Journal of Management, 11*(1), 1–15.

Buera, F., Fattal-Jaef, R., Neumeyer, A., & Shin, Y. (2020). *The economic ripple effects of COVID-19. Unpublished Manuscript*. Available at the World Bank Development Policy and COVID-19—ESeminar Series.

Calabrese, A., Costa, R., Menichini, T., Rosati, F., & Sanfelice, G. (2013). Turning corporate social responsibility-driven opportunities in competitive advantages: A two-dimensional model. *Knowledge and Process Management, 20*(1), 50–58.

Carroll, A. B., & Shabana, K. M. (2010). The business case for corporate social responsibility: A review of concepts, research and practice. *International Journal of Management Reviews, 12*(1), 85–105.

Conner, K. R. (1991). A historical comparison of resource-based theory and five schools of thought within industrial organization economics: Do we have a new theory of the firm? *Journal of Management, 17*(1), 121–154.

D'Aprile, G., & Mannarini, T. (2012). Corporate social responsibility: A psychosocial multidimensional construct. *Journal of Global Responsibility, 3*(1), 48–65. https://doi.org/10.1108/20412561211219283

Dey, M., & Sircar, S. (2012). Integrating corporate social responsibility initiatives with business strategy: A study of some Indian companies. *IUP Journal of Corporate Governance, 11*(1), 36.

Diamond, C. P. (1996). *Environmental management system demonstration project: Final report*. NSF, Inc.

Doering, D. S. (2002). *Tomorrow's markets: Global trends and their implications for business*. World Resources Institute.

Elkington, J. (1998). Accounting for the triple bottom line. *Measuring Business Excellence, 2*(3), 18–22.

Gentilini, U., Almenfi, M., Orton, I., & Dale, P. (2020). Social Protection and Jobs Responses to COVID-19.

Grzesiuk, K. (2017). Communicating a company's CSR activities through social networks: A theoretical framework. *Annales. Ethics in Economic Life, 20*(4), 89–104. https://doi.org/10.18778/1899-2226.20.4.07

Haropoulou, M. (2013). *Organizational decision-making and strategic product creation in the context of business sustainability outcomes: Theoretical synthesis and empirical findings* (Doctoral dissertation, Lincoln University).

Herzberg, F. (1964). *The motivation-hygiene concept and problems of manpower*. Personnel Administration.

Hishan, S. S., Ramakrishnan, S., & Jusoh, A. (2020). Corporate social responsibility for climate change using social contracts: A new research agenda. *Annals of Tropical Medicine and Public Health, 23*(6), 144–147. https://doi.org/10.36295/ASRO.2020.23618

Holliday, C. O., Schmidheiny, S., & Watts, P. (2002). *Walking the talk: The business case for sustainable development*. Berrett-Koehler Publishers.

Knox, S., Maklan, S., & French, P. (2005). Corporate social responsibility: Exploring stakeholder relationships and programme reporting across leading FTSE companies. *Journal of Business Ethics, 61*(1), 7–28.

Kohlberg, L. (1985). *Kohlberg's stages of moral development* (pp. 118–136). WC Crain, Theories of Development.

Livingston, E., Desai, A., & Berkwits, M. (2020). Sourcing personal protective equipment during the COVID-19 pandemic. *JAMA, 323*(19), 1912–1914.

Macagno, T. (2013). A model for managing corporate sustainability: Business and society review. *Business and Society Review, 118*(2), 223–252. https://doi.org/10.1111/basr.12009

Martinuzzi, A., & Krumay, B. (2013). The good, the bad, and the successful—How corporate social responsibility leads to competitive advantage and organizational transformation. *Journal of Change Management, 13*(4), 424–443.

Maslow, A. H. (1943). Preface to motivation theory. *Psychosomatic Medicine, 5*, 85–92.

McQueen, D. (2019). Frack off: Climate change, CSR, citizen activism and the shaping of national energy policy. In *Responsible People* (pp. 175–198). Springer.

Mirvis, P. (2011). Chapter 2 Unilever's drive for sustainability and CSR—Changing the game. In S. Albers Mohrman & A. B. (Rami) Shani (Eds.), Organizing for sustainable effectiveness (pp. 41–72). Emerald Group Publishing Limited. https://doi.org/10.1108/S2045-0605(2011)000000 1007

Oginni, O., & Omojowo, A. (2015). The implementation of sustainable business model among industries in Cameroon. *OIDA International Journal of Sustainable Development, 8*(11), 71–80.

Palazzo, G., & Richter, U. (2005). CSR business as usual? The case of the tobacco industry. *Journal of Business Ethics, 61*(4), 387–401.

Porter, M. E., & Kramer, M. R. (2006). Strategy and society: The link between competitive advantage and corporate social responsibility. *Harvard Business Review, 84*, 78–92.

Puppim de Oliveira, J. A., & Jabbour, C. J. C. (2017). Environmental management, climate change, CSR, and governance in clusters of small firms in developing countries: Toward an integrated analytical framework. *Business & Society, 56*(1), 130–151. https://doi.org/10.1177/0007650315575470

Ray, S. (2013). Linking public sector Corporate Social Responsibility with sustainable development: Lessons from India. *RAM. Revista De Administração Mackenzie, 14*(6), 112–131. https://doi.org/10.1590/S1678-697120 13000600006

Robinson, S., Smith, J., D'Aprile, G., & Mannarini, T. (2012). Corporate social responsibility: A psychosocial multidimensional construct. *Journal of Global Responsibility, 3*(1), 48–65.

Rupp, D. E., Skarlicki, D., & Shao, R. (2013). The psychology of corporate social responsibility and humanitarian work: A person-centric perspective. *Industrial and Organizational Psychology: Perspectives on Science and Practice, 6*, 361–368. https://doi.org/10.1111/iops.12068

Schmidheiny, S., & Timberlake, L. (1992). *Changing course: A global business perspective on development and the environment* (Vol. 1). MIT Press.

Schönherr, N., Findler, F., & Martinuzzi, A. (2017). Exploring the interface of CSR and the Sustainable Development Goals. *Transnational Corporations, 24*(3), 33–47. https://doi.org/10.18356/cfb5b8b6-en

Stern, V., Peters, S., & Bakhshi, V. (2010). *The stern review*. Government Equalities Office, Home Office.

Visser, W. (2010). The age of responsibility: CSR 2.0 and the new DNA of business. *Journal of Business Systems, Governance and Ethics, 5*(3), 7.

WCED, S. W. S. (1987). World commission on environment and development. *Our Common Future, 17*, 1–91

Williamson, V., Murphy, D., & Greenberg, N. (2020). COVID-19 and experiences of moral injury in front-line key workers. *Occupational Medicine, 70*(5), 317–319.

CHAPTER 6

Behavioural Transformation for Sustainability and Pro-Climate Action

INTRODUCTION

At one side, humans are significantly responsible for causing the change in climate and at another side, humans are the ones going to face the negative impact of climate change (Gifford et al., 2011). Hence, human behaviour and lifestyle leaves long-term cumulative influences on sustainability of ecological systems (Verplanken, 2018). Unsustainable activities such as overconsumption of resources resulting from human overpopulation; deforestation leading to impacts on biodiversity; natural habitat loss for flora and fauna and extinction of rare species of plant and animals; industrial activities from coal, mining, nuclear and other industries leading to environmental degradation; use of non-recyclable plastics, harmful pesticides; and increasing concrete jungles resulting from unmindful urbanization, have all taken a heavy toll upon the physical health and well-being of the planet, contributing to global climate change. The way people across the globe have contributed to climate change through their unsustainable behavioural practices, climate change has primarily become a concern of psychological and behavioural sciences, especially pertaining to the field of environmental psychology (Schmuck & Schultz, 2012). Therefore, climate crisis across the globe cannot just be limited to explore climate solutions at the levels of corporate bodies and governance systems. 'Solving the global climate change crisis is going to rely on, in one way or

© The Author(s), under exclusive license to Springer Nature Singapore Pte Ltd. 2022
P. Rishi, *Managing Climate Change and Sustainability through Behavioural Transformation*, Sustainable Development Goals Series, https://doi.org/10.1007/978-981-16-8519-4_6

another, changing human behaviour' (Williamson et al., 2018). Hence, we are looking forward to behaviour-centric solutions, at the level of individual, family, and community, which are the starting points for any behavioural change to happen.

Along with climate change, sustainability initiatives can also be made more effective by expanding our knowledge and understanding of human being and their behavioural practices. Merely knowing about sustainability is not enough. There is a need to bridge the gap between 'theory and practice' in the domain of pro-sustainability behaviours. Rather application of knowledge to alter and modify our existing unsustainable behaviours to make them eco-friendlier is needed. The review of mainstream climate change research reveals overwhelming importance to technological solutions for climate action paying considerably limited attention to pro-climate behavioural transformation (van de Ven et al., 2018). In the similar context, Stoknes aptly remarks that we need to shift from 'talking about the climate system to talking about people's responses to climate science'.[1] Hence, social and behavioural response to climate change is as pertinent as the climate change itself (Stoknes, 2014) and there is a need for systematic behavioural interventions for implementing climate change solutions. Multidisciplinary approaches integrating natural and social sciences are warranted in this regard (Gifford et al., 2011).

The Concept of Behavioural Transformation

According to IPCC (2012), Transformation is explained as 'physical and/or qualitative changes in form, structure, or meaning making', or as 'the altering of fundamental attributes of a system (including value systems; regulatory, legislative, or bureaucratic regimes; financial institutions; and technological or biological systems)'.[2] Further, it is also

---

[1] https://e360.yale.edu/features/how_can_we_make_people_care_about_climate_change.

[2] IPCC (2012) Glossary of terms. In C.B. Field, V. Barros, T.F. Stocker, D. Qin, D.J. Dokken, K.L. Ebi, M.D. Mastrandrea, K.J. Mach, G.-K. Plattner, S.K. Allen, M. Tignor, and P.M. Midgley (eds) Managing the Risks of Extreme Events and Disasters to Advance Climate Change Adaptation A Special Report of Working Groups I and II of the Intergovernmental Panel on Climate Change (IPCC). Cambridge, UK, and New York, NY, USA: Cambridge University Press: 555–564. *(13) (PDF) Responding to climate change: The three spheres of transformation.* Available from: https://www.researchgate.net/publication/309384186_Responding_to_climate_change_The_three_spheres_of_transformation [accessed Sep 22 2021].

explained as a psycho-social process through which human potential for commitment to change for a better quality of life is unleashed in order to promote long-term changes for the betterment in relation to self, others and the planet as a whole (Schlitz et al., 2010). Considering the context of climate change, 'transformation is a complex process that entails changes at the personal, cultural, organizational, institutional and systems levels' (O'Brien & Sygna, 2013).

Transformation broadly looks for incremental societal and behavioural change through cooperation, trust, and learning, in response to ecological crisis. Role of political, cultural, and economic institutions along with leadership and power relations are also crucial to make it happen (Brand, 2016). The terminology of transformation is increasingly being discussed amongst social and behavioural scientists and policy experts engaged in climate change adaptation and sustainability research as reflected in the fifth assessment report of the Intergovernmental Panel on Climate Change (IPCC, 2014).[3] Behavioural transformation is possible through a systematic and planned strategy to promote incremental changes in existing behavioural practices of people, taking into consideration the socio-cultural context and individual behavioural choices.

While recommending various approaches to manage risks through climate change adaptation, a threefold transformation strategy (3Ps) was suggested in IPCC fifth assessment report which is based on Sharma (2007).[4] Firstly, a *Practical* approach integrating socio-technical innovations with behavioural shifts to bring out positive outcomes. Secondly, a *Political* approach, reducing climate change vulnerability and risk in order to promote adaptation, mitigation, and sustainable behavioural practices through political, socio-cultural, and ecological decisions and actions. Thirdly, a *Personal* or individual-centric approach working on 'individual/collective' belief systems, values, and practices governing climate change responses. Besides recommending the need for adaptation efforts on the part of individuals, communities, and societies, sometimes there may be need for 'transformational adaptation' which goes 'beyond incremental approaches to climate change impacts, and may include changes...through novel, large-scale actions....in anticipation of, or in response to observed or expected impacts, involving coordinated or

---

[3] https://www.ipcc.ch/site/assets/uploads/2018/05/SYR_AR5_FINAL_full_wcover.pdf.

[4] https://www.kosmosjournal.org/article/personal-to-planetary-transformation/.

uncoordinated actions……deliberate or inadvertent (Park et al., 2012). While describing the types of transformational adaptations, Kates et al. (2012) classifies them as 'those adopted at a larger scale or intensity; those that are novel to a particular region or system; and those that transform places or involve a shift in location'. Further, 'although transformational adaptations are most often technological or behavioural, it is recognized that there are legal, social and institutional barriers linked to values, ingrained behaviours, and self-identities' (Kates et al., 2012).

IPCC also laid emphasis on transformational adaptation with the introduction of climate-friendly technological practices and governance systems facilitating adaptation at various levels in an equitable and ethical manner. To establish its efficacy at the national level, transformation must reflect visions and approaches of the country to achieve SDGs in line with its context and priorities.

IPCC sixth assessment report (2021) emphatically states 'The evidence of human influence on the climate system has strengthened progressively over the course of previous five IPCC assessments'.[5] Climate change and sustainability essentially warrants the transformation of human behaviour. How human beings process, respond to, and disseminate information and identify the enabling factors to transform 'awareness to action, and action to sustained behaviour change' comes under the purview of behavioural science to facilitate people in shifting their behavioural choices in favour of sustainable practices. UN Environment climate change expert Niklas Hagelberg remarks 'People in general are positive to climate change and carbon neutrality but these can be abstract concepts and remote to many people's daily lives….Behavioural science and behaviour change approaches are therefore critical to support shifts in behaviour'.[6] The behavioural and cultural aspects of climate change (Swim et al., 2011) and associated psychological barriers to pro-climate action have already been reported by Gifford et al. (2011). Cognitive psychological analysis narrates some of the mental strategies which operate when people are dealing with the adversities associated with climate change and trying to build up their possible response (Kahan, 2012). However, as climate change impacts not only individuals but societies and cultures at large, transformational efforts must extend beyond people, to societies and

---

[5] https://www.ipcc.ch/report/ar6/wg1/downloads/report/IPCC_AR6_WGI_Chapter_03.pdf.

[6] https://www.unep.org/news-and-stories/story/five-ways-behavioural-science-can-transform-climate-change-action.

cultures too, to have a larger impact. This aspect is being analysed in the forthcoming section.

Behavioural Transformation Across Societies and Culture

Complex planetary challenges require more than the technological or policy-oriented interventions making it clear that sociocultural practices are pertinent in the upcoming times and there is a need for 'cultural shift towards sustainability-linked climate change adaptation' (Rishi & Schleyer-Lindenmann, 2020). A significant role of cultural values has already been described by (Crompton, 2011) in determining collective actions of people which comes under the purview of social psychology. Norgaard (2011) analyses the sustenance of climate change scenario 'through the social construction of denial, and through the cultural management of emotions'. Hence, culture has gained attention as one of the significant mediating variables in the human–climate change connect, to establish how societies perceive, responds to and act on climate change scenario (Adger et al., 2013; Rishi & Schleyer-Lindenmann, 2020). Climate change and associated intention for pro-climate action is diversely connected to national cultures and demographics. Individuals and societies, on the whole, vary in their adaptive strategies to climate change as well as climate action as a function of situational or personal variables. Individual values and belief systems coupled with cultural practices, which are not in line with mitigation efforts, fail to succeed and do not derive any significant behavioural change (Dietz et al., 2007). Gifford et al. (2011) also highlights 'social and political ideologies and beliefs as psychological barriers to climate change mitigation and adaptation practices'. For the regions like Western Europe and the United States, with the limited population, sufficient financial resources, and availability of technological innovations to create mitigating infrastructure, the adaptive capacity to withstand the short-term impacts of disruptive weather systems/other climate disasters is relatively high. However, southern Europe, Latin America and most of the South Asian countries like Indonesia, Thailand, Philippines, or Bangladesh, have high concern for climate change associated with readiness for pro-climate action, due to their higher vulnerability to climatic events, adversely affecting their day-to-day life.[7]

---

[7] https://e360.yale.edu/features/how_can_we_make_people_care_about_climate_change.

Therefore, it is well evident that efforts for mitigation, adaptation as well as climate action should be in line with the socio-cultural context of respective region, in order to ensure their long-term success. In the regions, where there is a visible mismatch between them, dedicated and systematic education/awareness generation efforts to enhance climate/sustainability consciousness (explaining how their existing beliefs are counteracting with natural environment and impacting their community and society at large) must precede mitigation/adaptation efforts. Community-based adaptive management of natural resources have the potential to enhance the resilience to climate change (Tompkins & Adger, 2004) and also to help people better engage with climate action at the local level.

An Interplay of Behavioural Dynamics

It is firmly acclaimed by now that climate change is a global environmental problem requiring a solution-centric approach at the execution level. To make it happen, there is a need for implementation of mammoth transformative efforts at multiple levels on the part of corporate organizations, government, and societies at large across nations. But undoubtedly 'individuals lie at the heart of true societal transformations'. We have learned about how human mindset and thought processes influence actions driven towards climate change and sustainability. We have also discussed the role of education, awareness, and communication in this regard. Behavioural change falls under the discipline of behavioural science in which a range of change models have been brought out to elucidate the change process. With climate change and its consequences, believed to be among the most vital challenges for humanity and the Earth's ecosystem, it is important to understand why individuals do or do not adopt pro-environmental attitudes and behaviours. Behavioural dynamics, especially the personality traits are well suited for this purpose (Soutter et al., 2020).

An interplay of various personality attributes is responsible for people to remain motivated/not bothering for sustainability and climate action. One such attribute is the tendency to procrastinate which is labelled as 'temporal self–regulation failure reflecting a disjunction between the present and future self'. It is also reflected through a limited future orientation (Sirois, 2014). While thinking to take action for any issue, which requires their significant attention, procrastinators try to postpone it for some future time till it becomes urgent. Ersner-Hershfield et al.

(2009) called it 'future-self continuity,' which is a differential perception of people towards present self and future-self in regard to their concern for future. While applying this hypothesis to climate action, it can be proposed that more detachment from future-self might be associated with less likelihood to go for climate action through behavioural change, if they do not find any short-term future benefit associated with it.[8] The cost of this behavioural tendency may range from billions to trillions of any currency and as the period of procrastination increases the regrets increase which may gradually close the opportunity to limit rising temperatures across the globe (Keller et al., 2007). Hammonds (2020) calls this tendency as a step towards global violation of human rights threatening humanity. He further calls for 'a transformative approach to our fundamental existence, including what we eat, how we live, and how we travel and commute' ......and recommends 'to push for transformative, progressive, rights-based engagement with the climate emergency' which, in short, is a need for behavioural transformation for climate justice. Otherwise, we all will be woefully facing the irreversible climatic emergency at the global front with no one to be spared by its impacts.

In a webcast, '*Wired to Ignore Climate Change*' by George Marshall,[9] the importance of presenting information about climate change in a 'narratively satisfying' manner, appealing to our 'emotional brain' is highlighted. Such reflection of information can make climate change more personally relatable through cultivation of positive emotions which have better likelihood for transformative behaviour and climate actions. On the other hand, threat producing negative climatic imagery and news generating negative emotions generally push for avoidance and procrastination at the behavioural front and make people passive and indifferent to the issue of concern.

A good deal of recent research is increasingly focusing on understanding the link between various personality factors/traits and pro-environmental behaviour leading to climate action and sustainability. Ribeiro et al. (2016) studied sustainable consumption from the consumer's point of view. More specifically, the authors investigated which personality traits could be antecedents of sustainable consumption behaviour. The findings revealed that 'frugality' and 'conscientiousness'

---

[8] https://climate.org/individuals-and-climate-change-facilitating-behavior-change-for-societal-transformation/.

[9] https://www.youtube.com/watch?v=5cw710DgM1s.

traits are those that have the greatest relationship with the trait propensity to sustainable consumption behaviour, represented by the preference for purchasing ecological products, saving resources and carrying out the recycling of materials.

Further, Locus of Control (LOC), also known as attributional style in behavioural science literature, happens to be an important factor which can predict climate action. Research suggests that LOC, or the extent to which a person feels his or her own actions can produce broader change, is an important predictor of environmental behaviour (Giefer et al., 2019). Locus of control is of two types: Internal Locus of Control (ILOC) and External Locus of Control (ELOC). While ILOC denotes individuals' 'multifaceted attitudes pertaining to personal responsibility towards and ability to affect environmental outcomes', ELOC on the other hand 'encapsulates attitudes towards environmental outcomes that individuals believe are the result of extraneous forces beyond their volition' (Cleveland & Kalamas, 2014). Various studies have tried to study the relationship between LOC and pro-environment behaviour and associated climate action. Giefer et al. (2019) in their study have used a nationwide survey from China to test whether LOC moderates the effect of environmental attitudes on behaviour. The results revealed that respondents with ELOC (i.e. those who believe personal actions cannot produce change) reported lower pro-environmental behaviour than those with ILOC (i.e. those who believe personal actions can produce change). Chiang et al. (2019) have also highlighted LOC as a crucial factor in pro-environmental behaviour. In their study, the authors found that 'emotional stability' may be a predisposing factor for internal locus of control and pro-environmental behaviour and hence people with higher emotional stability and a stronger internal locus of control are more likely to engage in pro-environmental behaviour which can be one of the indicators of sustainable behaviour and climate action too.

Various other personality traits have also been studied in the context of sustainability, pro-environment behaviour, and climate action. Rothermich et al. (2021) have used a trait-level approach in their study to understand pro-environmental behaviour in the context of climate change. The study tested the correlation between self-reported Big Five traits, trait-level anxiety, and empathy and found that Openness, Perspective Taking, sex, and age correlate with climate change attitudes. Such studies increase our understanding of environmental challenges to the general public and offer implications for future research on how to execute pro-environmental strategies for promoting climate action. Akbar et al. (2020)

have also analysed the relationship between the big five personality traits and sustainable consumption behaviour among social networking sites users from Pakistan and found a positive association between the two.

In another study by Panno et al. (n.d.), the authors investigated whether and how HEXACO personality traits relate to climate change action, pro-environmental behaviour, and moral anger. The HEXACO model of personality structure is a six-dimensional model of human personality created by Ashton and Lee includes factors like Honesty-Humility (H), Emotionality (E), Extraversion (X), Agreeableness (A), Conscientiousness (C), and Openness to Experience (O).[10] The study by Panno et al. (n.d.) indicated that (1) Openness to Experience outperformed the other HEXACO personality traits in predicting climate change action, whereas (2) both Openness to Experience and Honesty–Humility outperformed the other HEXACO personality traits in predicting pro-environmental behaviour. The study also revealed that that (3) Openness to Experience was related to climate change action both directly and indirectly via moral anger. Furthermore, (4) Openness to Experience and Honesty–Humility were independently related to pro-environmental behaviour both directly and indirectly via moral anger. Studies like these have important implications for understanding the role of personality traits in climate action. Soutter et al. (2020) also performed a meta-analysis of the associations of the Big Five and HEXACO personality domains with pro-environmental attitudes and behaviours and found 'openness' and 'honesty-humility' as the strongest correlates of pro-environmental attitudes and behaviours.

Sussman et al. (2016), in their report on role of social mobilization in promoting climate action, explains some psychological/behavioural models to elucidate motivations to approach or avoid climate action like rational choice (Scott, 2000) and altruistic approaches (Schwartz, 1977). Besides, educational models, intrinsic-extrinsic motivation models (Ryan & Deci, 2000), norm activation models (Schwartz, 1977), and values-attitude-behaviour models (Stern et al., 1999) also try to contribute towards promoting behavioural shifts in different ways.

No education is complete until and unless it is able to create observable behavioural impacts. Sustainability education emerged as one of the foremost strategies to impact pro-climate shift in behavioural practices.

---

[10] https://hexaco.org/#:~:text=Here%20you%20will%20find%20basic,Agreeableness%20(versus%20Anger).

It can enable people to take informed decisions about climate change and associated issues (Boyes & Stanisstreet, 2012). That is why, education remains one of the primary strategies in various climate change campaigns. However, it is most effective when combined with other strategies rather than working in isolation. One of the early models in this regard was by (Ramsey & Rickson, 1976). They proposed that imparting knowledge about environmental issues through systematically designed environmental education programmes can enhance environmental awareness and cause attitude change, subsequently leading to behavioural change. Further, Hines et al. (1987) also emphasized role of knowledge, but stated that it cannot work in isolation. Psycho-social factors, behavioural intentions, and contextual/situational factors also play a significant role in promoting desirable behavioural shifts. The detailed explanation of sustainability education and communication has already been elaborated in previous chapter. Many time, use of reinforcement/extrinsic motivation can also be effective to initiate the behavioural shift towards environmentally desirable behaviour, especially in younger population.

Mindset is the potent mode to promote intentional and incremental changes in pro-climate behaviour which starts at individual level and expands beyond community to societal and global level representing larger shift in social and behavioural norms. A strong level of climate-centric commitment can reflect huge differences across societies. However, every possible transformation in behavioural practices leading to pro-climate action warrants a mindful approach on the part of people at the level of Intention, Attention and Action (IAA) (Fig. 6.1).

In means that people go for any possible behavioural shift only when they intend to do so. Transformation process must start with creating

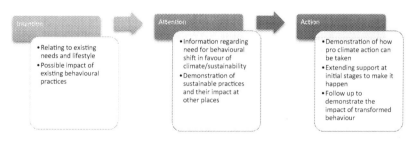

Fig. 6.1 IAA framework for behavioural transformation

intention, which is possible only when the issue relates closely to their existing needs and lifestyle and they are able to understand the impact of their existing behavioural practices on climate and environment in general. Further, it may be coupled with information regarding possible shifting of behaviour in favour of climate/sustainability, to which they may pay attention. Their attention will be more focussed when they are able to observe themselves how others are demonstrating sustainable practices. When these two steps of intention and attention are crossed, possibility of associated action and likely transformed behaviour (pro-climate action/sustainable behavioural practices) becomes high. At that stage, people should be supported to practice pro-climate action at micro-level with initial support and then follow up until behavioural shift is established. In the ideal sense, the model appears quite logical to happen but in practice, human behaviour is as complex as the phenomenon of climate change. While making efforts for implementation of IAA framework, many times, in spite of intention and attention to desirable behavioural cues, behavioural transformation does not happen/sustain for long. The underlying process may be awareness-action/attitude-behaviour inconsistency as described below.

Awareness-Action Inconsistency—Role of Emotions
In this era of digital information, there may be hardly any dearth of information in regard to the phenomenon of climate change and its impacts. Despite this heightened awareness, which is the result of 'climate talk' across the globe, the indifference prevails in people for taking pro-climate action or as far as 'climate walk' is concerned. Its primary reasons may be the fact that most of the awareness programmes pertaining to climate/environmental sustainability disseminate the facts and figures only while laying little emphasis on making people understand the impact of those facts and figures on them and what can be done by people at an individual level to make a difference in emerging climate change scenario (Wi & Chang, 2019). As a result, at the affective level, most of the people show considerable worry about the likely impacts of climate change. Recently, IPCC (2021),[11] sixth assessment report also reflected

---

[11] IPCC, 2021: Climate Change 2021: The Physical Science Basis. Contribution of Working Group I to the Sixth Assessment Report of the Intergovernmental Panel on Climate Change [Masson-Delmotte, V., P. Zhai, A. Pirani, S. L. Connors, C. Péan, S. Berger, N. Caud, Y. Chen, L. Goldfarb, M. I. Gomis, M. Huang, K. Leitzell, E. Lonnoy,

fear-producing facts and figures of sea-level rise due to glacier melting, resulting in submerging of 12 coastal cities of India including Mumbai, Chennai, Kochi, Vishakhapatnam, Bhavnagar, Kandla port, Mangalore and many others creating a fear appeal among people. But the information regarding concerted efforts to be taken from local to global level including individual-level action points are rarely highlighted in media reports to draw significant public attention. As a result, in the absence of concrete climate action points and the tendency of people to push all painful and unpleasant information out of their conscious mind (Freud & Bonaparte, 1954), the outcome of this worry, in regard to motivating specific pro-climate action at individual level remains largely unknown (Bouman et al., 2020). In the similar context, Stoknes remarks, 'if you overuse fear-inducing imagery, what you get is fear and guilt in people, and this makes people more passive, which counteracts engagement'. Therefore, dissemination of research and information 'in an opportunistic manner to raise moods and induce creativity to find solutions' may give better results. He further calls for reframing of discussion on climate change stating... 'We need a new kind of stories....that tell us that nature is resilient and can rebound and get back to a healthier state, if we give it a chance to do so'.[12]

Shiota et al. (2021) also highlighted the importance of affective aspect of behaviour, particularly emotions, in motivating people for behaviour change. Contradicting with the previous research on integrating fear appeals for behavioural interventions, recent theoretical/empirical evidences suggest that 'positive affect and emotions can promote change by serving as proximal rewards for desired behaviours'. Hence, integration of positive affect and emotions while planning and implementing behavioural interventions for climate change and sustainability are recommended. Going deeper into the interrelationship between thought processes and climate action, one of the most common observations is that many of the times we have clear understanding of the likely negative impacts of climate change or unsustainable behavioural practices but we still refrain from taking desirable actions at the behavioural level. Why it is so? Why we understand and still can't turn

---

J. B. R. Matthews, T. K. Maycock, T. Waterfield, O. Yelekçi, R. Yu and B. Zhou (eds.)]. Cambridge University Press. In Press.

[12] https://e360.yale.edu/features/how_can_we_make_people_care_about_climate_change.

it into practice? Interventions for changing behaviour generally extend temporary support and many times such behavioural change does not persist for long and have limited efficacy.

While examining the factors associated with initiation and maintenance of behaviour change, some of the significant emerging concepts are individual habits and motivations, self-regulation, availability of physical and psychological resources and external environmental and social influences (Kwasnicka et al., 2016). In the absence of a strong psychological framework pushing people towards action, with the influence of some pushing force in external environment, the desirable behaviour may initiate temporarily but its maintenance/continuance over time is not assured. Besides, while taking conscious or unconscious judgements/decisions, human beings have a tendency to deviate from rationality and resort to cognitive short-cuts to reach an acceptable outcome (Gigerenzer & Gaissmaier, 2011) which are named as self-serving probability biases. It is a common tendency of human beings to largely overestimate the likelihood of positive outcomes of all events happening to them and their near and dear ones while underestimating the likelihood of all negative events like climate change linked disasters/impacts, happening in their region. This tendency adversely acts on preparedness and makes people helpless at the time when emergency action in response to climatic events is warranted.[13] The common response is that we read a lot about climate change but we never thought it will actually happen in my region too and we have to take some action.

While understanding the issues concerning implementation of sustainability objectives, Blake (1999) reflected the need for differentiated policies, taking into consideration the socio-economic and political scenario. It can address the 'value-action gap' and promote local partnerships and participation while being sensitive to local diversity, so that equitable distribution of responsibility among stakeholders can be ensured.

BEHAVIOURAL TRANSFORMATION FOR SUSTAINABILITY

Sustainability encompasses striking a judicious balance between the three P's i.e. people, planet, and profit. Sustainability of present and future generations on this planet earth depends, to a great extent, on our

[13] https://climate.org/individuals-and-climate-change-facilitating-behavior-change-for-societal-transformation/.

current and future behaviours. There is not much that we can do now to undo the damage that has already been caused because of thoughtless human behaviour. But we can certainly contribute our bit to ensure that we do not engage anymore in unsustainable behaviours which are against the planet's well-being. Fischer et al. (2012) mentions that 'sustainability demands changes in human behaviour'. He also recommended some of the priority areas of change like 'reforming formal institutions, strengthening the institutions of civil society, improving citizen engagement, curbing consumption and population growth, addressing social justice issues, and reflecting on value and belief systems'. So, changing our old habits, lifestyles and adopting planet-friendly behaviours is crucial for promotion of sustainability behaviour. Any change is not easy in the beginning, especially when it comes to changing our habits. The process comes with its own set of barriers that make it difficult to break old and unsustainable practices. The barriers to sustainability no longer rest in limited awareness/knowledge about arising biophysical and social issues due to unsustainable practices. The challenge lies in how to walk on the path of sustainability and make our active contributions for sustainable future (Fischer et al., 2007). However, there is an urgent need to reform our consumption practices so as to make them more and more aligned with sustainability.

Sustainable Consumption Behaviour

'Consumption' and 'delivery' are the two ends of the sustainability continuum. To make both these ends meet should be the goal of sustainable human behaviour. It is not only that we consume responsibly but it is also our duty to deliver back what we consumed and that too in a responsible manner. Sustainable consumption behaviour (SCB) originates with a concern for environmental well-being and is aimed to provide a better quality of life for our upcoming generations. The recognition of the need to provide a safe and liveable planet to future generations is at the base of SCB. It also stems from a 'behaviour intention' to engage in such behaviour. Without intention or willingness, SCB cannot take place. This behaviour intention, in turn, has its roots in several factors (Wang et al., 2014).

First, a person's environmental values go a long way in motivating someone for the decision to engage (or not) in SCB. Values are an integral part of one's personality, especially if such values are strongly held as opposed to loosely held values. Largely, our environmental attitudes have a moderating impact upon our environmental values. SCB further depends upon whether these values are internally oriented or externally oriented. In most cases, internally oriented values tend to have a stronger impact upon SCBs as compared to environmental values with an external orientation.

Second, knowledge about the environment can give people insight into the existing state of affairs and what is happening to their environment. There are good chances that the more knowledge a person has about environmental issues, the more he or she is likely to engage in SCB although this might not hold true always. Third, wanting to take responsibility for the protection and maintenance of environment can be a major determinant of SCB. When a person chooses pro-environmental behaviour by choice, he contributes more towards sustainability. Such voluntary actions are not only long-lasting but also have more profound impacts upon environment health and well-being. Fourth contributing factor to SCB is sensitivity towards the environment. Having the required compassion and kindness can prompt individuals for environmental action. Fifth, perceived behavioural control is another major factor as to when a person feels that he can easily have the required control over his behaviour to make his contributions to sustainability, there are more chances of people engaging in meaningful SCBs as opposed to the situation when they feel lack of confidence and control over their behaviours.

Finally, the ability to visualize consequences of engaging (or not) in SCBs and the related perception of the whole set of advantages and disadvantages of such behavioural intentions can help people choose wisely and make decisions with regard to their consumption practices.

Two different factors operate when it comes to engaging the consumers in SCBs (Piligrimienė et al., 2020). First, internal factors which might include a person's attitude towards the environment, perceived responsibility and perceived behavioural efficacy and second, external factors like conditions conducive to SCB and also the social environment or surroundings of an individual. These factors impact an individual on three different levels viz. the cognitive level (thoughts and higher mental processes), affective level (emotions), and the conative level (actions and behaviours). Examples include (i) the behaviour encompassing awareness about responsible consumption and conservation of resources (at cognitive level), (ii) human belief systems and values (at affective level), and (iii) adopting sustainable practices in life (at the conative level).

White et al. (2019) have discussed a comprehensive framework for conceptualizing and encouraging sustainable consumer behaviour change, which is represented by the acronym SHIFT. It proposes that consumers are more inclined to engage in pro-environmental behaviours when the message or context leverages psychological factors like Social influence (S), Habit formation (H), Individual self (I), Feelings and cognition (F), and Tangibility (T).

In present times, a number of recent trends have emerged for sustainable living which are gaining momentum and popularity among individuals owing to their several advantages. Plogging is one such sport activity which originated in Sweden. The word combines 'running' and the Swedish expression 'plockaupp', which means 'picking up' and includes a combination of jogging and picking up trash that one comes across along the way at the same time. Plogging allows one to enjoy the dual benefits of health and contribution to environment at the same time. Upcycling is another concept which is creative reuse of old objects to create new products. It serves one's creativity and imaginations while at the same time is an eco-friendly activity too.[14]

Next, pre-cycling is another activity which aims at reducing waste from the time of purchase itself. It emphasizes upon buying items that are needed and that will last, discourages excess packaging, and also points at reusing as much as possible. Pre-cycling chiefly aims at encouraging to buy products which have least impact on the environment.

A step ahead is the 'Zero Waste' movement which is based on the five error rule: reject what you don't need, reduce what you do need, reuse packaging and materials and opt for second-hand consumption, recycle everything for which you cannot stop, or reduce your consumption or which cannot be decomposed to obtain natural fertiliser (Fig. 6.2).

Opting for second-hand items is being increasingly adopted by many people these days owing to the several benefits associated with this practice. Every time a person chooses to buy a used item instead of something new, not only do they save money but it also means they aren't releasing emissions into the environment and are contributing to sustainable development. Lastly, sustainable mobility is also an area which has received much attention. This is implemented in the form of preference

[14] https://www.activesustainability.com/sustainable-life/seven-trends-to-follow-to-live-more-sustainably/.

Fig. 6.2 Desirable behavioural actions for sustainable living

for walking, use of a bicycle, using Public transport, bike-sharing, car-pooling, etc. as these are not only sustainable but have various benefits associated with them.

To sum up, we can say that 'sustainability is a normative concept, meaning it embodies a particular set of values' (Fischer et al., 2012). To move on sustainability walk, there is a need for effective transdisciplinary sustainability science, recognizing the significant role of individuals and civil society, and going for their active engagement in sustainability initiatives. It requires the significant role of societal change agents as well as participation of indigenous people. Change agents have to make them understand the need to go beyond their comfort zones and set behavioural norms, which they have been practicing since ages, in order to proceed on the path of sustainability walk. Interdisciplinary collaboration and supportive local institutional context plays a crucial role to make it happen. Fischer et al. (2012), very appropriately remarks that 'a social avalanche is needed'. So the challenge is not the knowledge and awareness but to get started on sustainability walk, overcoming the barriers of old habits, traditions, behavioural norms and unsupportive institutional context. With a strong commitment for protecting the planetary health, it is definitely possible and will take place one day, no matter how challenging it appears as we are not left with any option but to behave sustainably.

Behavioural Transformation for Pro-Climate Action

As rightly remarked by Nicholas and Wynes, '... large-scale behavioural change, away from today's high-emitting lifestyles, is a prerequisite for meeting our Paris climate commitments'.[15]

Page and Page (2014) presented an Internal Transformation framework which can possibly underpin behaviour change for pro-climate action. Special Report of Intergovernmental Panel on Climate Change (IPCC) on 1.5 degrees Celsius (°C) categorically mentions climate change as 'the act of changing one's behaviour'. It further mentions that habit-laden individuals are posing additional challenges to address climate change. Therefore, there is a need for 'greener behavioural changes' by asking for 'environmentally-conscious alternatives' and integrating 'behavioural and lifestyle changes to be able to limit warming to 1.5 °C'.[16]

Building a basis for sustainable environment requires a greater understanding, anticipation, and preparation for probable climate change impacts. In doing so, the required action uptake is contingent on individuals' ability and willingness to modify their behaviour, which is influenced by a variety of factors. These dynamic factors, such as cognitive, socio-economic, political, and cultural factors, shape and determine behaviour, and are thus referred to as behavioural determinants.[17] Drivers of behaviour or behavioural determinants are specific for each behaviour but some of the typical behavioural factors related to determining climate adaption behaviour include perceived risk; perceived social norms; perceived self-efficacy; and perceived response efficacy. While perceived risk is the individual's or community's perceptions of the existence, probability, and characteristics of a given climate risk, perceived social norms comprise of how society perceives a specific action which can have a big impact on whether or not someone decides to do that climate action. Perceived self-efficacy is individual's belief in his or her own ability to adapt against climate change impacts and perceived response efficacy

---

[15] https://www.wri.org/climate/expert-perspective/changing-behavior-help-meet-long-term-climate-targets.

[16] https://www.downtoearth.org.in/blog/climate-change/ipcc-report-we-need-behavioural-change-not-climate-change-61843.

[17] http://repo.floodalliance.net/jspui/handle/44111/3060.

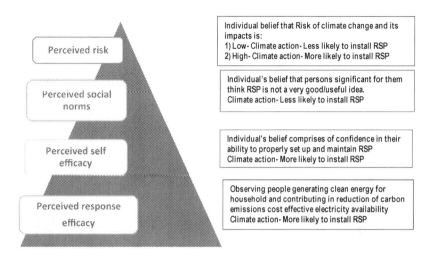

Fig. 6.3 Role of behavioural determinants in shaping specific behaviour

stands for individual's perception of a climate response's ability to minimize a climate risk or accomplish a goal. To have a better understanding on how these behavioural determinants shape a particular behaviour, the example of implementing Rooftop Solar Power (RSP) in a household as illustrated in Fig. 6.3 is worth to mention.

Figure 6.3 depicts the probable climate action that a person would take up based on given behavioural determinants that might shape up their decision and are thus crucial in incorporating behavioural changes in daily life. It can force us to think that in spite of behavioural intentions to transform for climate change and sustainability, there might be many barriers existing in our minds or society or community which may negatively impact the process of behavioural transformation as given in the subsequent section.

Climate Change Mitigation Through Behavioural Transformation

Mitigation and adaptation are two countermeasures that we can take in order to deal with climate change. To restrain global warming to a minimal, mitigation implies lowering greenhouse gas emissions and boosting greenhouse gas sinks (McCarthy et al., 2001). It is one of the most significant functions contributing to climate change, which comprises of scientific research, financial support, and technical practices (Zhao et al., 2018). Energy conservation through increased vehicle fuel efficiency; sequestration of carbon through tropical reforestation; and shifting to cleaner energy sources through changes in business operations are just a few of the essential technical measures. These methods, on the other hand, have proven to be incredibly slow and difficult to apply at best. Even if they do, due to the long lifetime of greenhouse gases in the atmosphere, their impact on global climate change will not be noticeable for decades (Pachauri & Reisinger, 2007). Furthermore, the existing impacts of climate change cannot be completely undone, underlining the adaptation as a necessity and urgency on both at individual and societal level (Ebi & Semenza, 2008; Haines et al., 2006; Zhao et al., 2018).

Adaptation refers to the regulatory approaches used in response to present or anticipated climatic stimulation, with the goal of reducing climate change impacts and increasing adaptive capability.

Apart from the aforementioned concept, a new component in this area has lately emerged, focusing on climate change mitigation and adaptation via the lens of behavioural change. A recent study in European Union emphasizes on the potential climate change mitigation through behavioural change comprising many behavioural options in food, mobility, and housing demand indicates that modest to rigorous behavioural change could reduce per capita footprint emissions by 6–16% (van de Ven et al., 2018). In another study, individual energy behaviour, in combination with social setting, determines energy demand, which leads to carbon emissions. In this manner, individuals can contribute significantly to a low-carbon route and global emission reduction through behavioural shifts (Niamir, 2019). Many models exist in the literature on individual choices for sustainable lifestyles, and one of them outlines three main factors for sustainable lifestyle choices: internal factors such as awareness, knowledge, values, attitudes, motivation, emotions, responsibilities, and priorities; internal factors such as cultural, social, economic, and institutional factors; and finally, demographic factors (Kollmuss & Agyeman, 2002). A study by Khanna et al. (2021) investigated interventions such as monetary incentives or feedback to encourage individuals for behavioural shift in the use of existing equipment in their households. Both monetary and non-monetary incentives' based interventions showed reduction in energy consumption of household and estimated a global carbon emissions' reduction potential of $GtCO_2$ yr^{-1} through deploying the right combinations of interventions. Another approach for climate change mitigation through behavioural change is offered by the aviation sector, in which airline passengers can actively choose airlines that are more environmentally friendly (Baumeister, 2020). Individuals in India are making conscious choices to keep their energy expenses within financial bounds and other possible interventions are mentioned in the Table illustrated below. In addition, there is a definite shift towards expanding and enhancing the usage of public transportation, which is, culturally and historically, a dominant mobility style prevalent there (Roy et al., 2018). The same has been widely practiced in Dayalbagh township of India where shared e-rickshaw is the primary mode of conveyance within the campus with extremely restricted entry of cars and two-wheelers running on fossil fuel. The township is rapidly moving ahead on the path of SDGs, adopting and practicing all possible modes of alternative energy with the technical and R&D support of scientific professionals of Dayalbagh Educational Institute and under the implementation guidance of visionary spiritual leader of Radhasoami faith (Box 6.1).

Box 6.1: Major recommendations for behavioural transformation

Frugal Choices
- Buy things only when necessary/ need replacement; Buy things only when necessary/ need replacement;
- Avoid shopping to stay trendy or upgraded Avoid shopping to stay trendy or upgraded
- Go for Frugal Innovation ('Jugaad') Go for Frugal Innovation ('Jugaad')

Food sustainability
- Choose carefully what you eat Choose carefully what you eat
- Buy local and organic Buy local and organic
- Avoid food wastage Avoid food wastage

Environment education and awareness
- Through inculcation of environmental values Through inculcation of environmental values
- Climate action and awareness initiatives Climate action and awareness initiatives

Environmental responsibility
- Reduce/avoid plastic/ non-recyclable products usage Reduce/avoid plastic/ non-recyclable products usage
- Avoid paper wastage Avoid paper wastage
- Carry reusable bottles, straws, cutlery etc Carry reusable bottles, straws, cutlery etc
- Reduce or avoid disposable items Reduce or avoid disposable items

(continued)

(continued)

Lifestyle choices	Adopt Upcoming Sustainability Trends
• Walk or use bicycle for shorter distances Walk or use bicycle for shorter distances • Use natural light and ventilation to replace avoidable air conditioning and artificial lighting Use natural light and ventilation to replace avoidable air conditioning and artificial lighting • Buy electrical appliances with higher star ratings Buy electrical appliances with higher star ratings • Conscious decisions to cut energy bill by using electricity responsibly • Adopting buildings with light coloured, more reflective rooftops[18] Adopting buildings with light coloured, more reflective rooftops • Opting for fuel efficient/electric cars Opting for fuel efficient/electric cars • Keeping fixed budget for fuel[19] Keeping fixed budget for fuel • Monitoring your carbon footprint Monitoring your carbon footprint • Conscious air travel choices • Composting Composting	• Plogging Plogging • Sustainable mobility (Car-pooling, ride sharing, electric and hybrid vehicles, etc.) Sustainable mobility (Car-pooling, ride sharing, electric and hybrid vehicles, etc.) • Zero waste movement Zero waste movement • Reuse, Precycle, Upcycle Reuse, Precycle, Upcycle • Stay @home sustainability Stay @home sustainability • Green and Sustainable housing Green and Sustainable housing • Adopting renewal energy (solar panels etc.) at household level[20] Adopting renewal energy (solar panels etc.) at household level • Rooftop organic gardening for green spaces Rooftop organic gardening for green spaces Take help from sustainability apps like: • Oroeco[21] Oroeco • Bikemap[22] Bikemap • Sustainable Development Goals • Love food hate waste[23] Love food hate waste • Refresh Go green[24] Refresh Go green • HowGood[25] HowGood • Know your carbon footprint[26] Know your carbon footprint

[18] http://shaktifoundation.in/wp-content/uploads/2014/02/cool-roofs%20manual.pdf.

[19] D. Chakravarty, Rebound Effect: Empirical Evidence from Indian Economy (Ph.D. thesis), Jadavpur University, India, 2015.

[20] http://www.ncpre.iitb.ac.in/research/pdf/Estimating_Rooftop_Solar_Potential_Greater_Mumbai.pdf.

[21] https://www.oroeco.com/.

[22] https://www.bikemap.net/.

[23] https://www.lovefoodhatewaste.com/.

[24] https://www.therefreshproject.com.au/go-green-app/.

[25] https://howgood.com/.

[26] https://www.knowyourcarbonfootprint.com/.

Barriers to Behavioural Transformation

Efforts for behavioural change are largely neglected by scientists and researchers as they are primarily perceived to be linked to values, beliefs, and cognitive attributes rather than to technological innovations and mitigation efforts which can be quantified and linked to climate models (Creutzig et al., 2016; Hardt et al., 2019). 'Psychological barriers also impede behavioural choices that would facilitate mitigation, adaptation, and environmental sustainability' remarks Gifford (2011). In his paper on 'The dragons of inaction', he further explains the need for interdisciplinary collaboration between psychologists, scientists, technologists, and policymakers so that civil society can be assisted to overcome these psychological barriers. Some of these barriers were listed as ignorance, limited cognition, ideologies, comparison with other people, discredance, perceived risk, and limited behaviour. Recently, Lacroix et al. (2019) developed 'the Dragons of Inaction Psychological Barrier (DIPB) scale' to measure the psychological barriers across six domains like 'food choices, transportation, energy use, water use, purchasing, and waste'. The scale can be adapted in climate change context to measure psychological barriers to climate change too in different demographic regions to compare and contrast and accordingly develop climate policies as well as adaptive strategies suiting to respective context.

Further, Stoknes (2015) has described 4D barriers to Climate action as given in Fig. 3.2. IPCC and other climate reports position their facts and figures in such a manner (far-reaching impacts by 2100) that it creates a psychological distance in the mind of people and the sense of urgency to act goes down. It acts as a barrier to climate communication efforts as perceived risk is not sustained for long to change behaviour. Further, presentation of doom and gloom phenomenon also acts as a barrier as most of the reports present climate change as a catastrophe creating fear-based imagery in the minds of people. They hardly try to use 'opportunity frame' to support people in visualizing what can be done on their part to avoid doom and gloom scenario (Fig. 6.4).

Denial stands for ignorance/negation/avoidance of facts concerning climate change which do not match with the long-existing behavioural practices and lifestyles of people. So, psychological denial works as an ego-defence mechanism to safeguard themselves from the anxiety that is generated within imagining the likelihood of catastrophic events associated with climate change. Lastly, dissonance associated with the inner

Fig. 6.4 4D barriers to climate action (based on Stoknes, 2015)

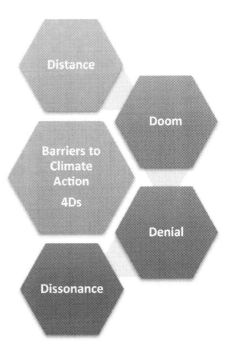

discomfort when there is a knowledge action gap.[27] For example, we know that fossil fuel consumption leads to global warming but we still continue to use them as the conveniently available source of energy, which might lead to dissonance. People use various ways to handle such dissonance and reduce associated anxiety like developmental needs of country like India cannot be disregarded due to low developmental indices or saying that anyway, the per capita carbon emission of country like India is very low so let developed countries first change their lifestyles, etc.

It is very clear from the available psychological literature that criticism of people refraining from climatically desirable behaviour will only lead to resistance to change. Hence, for overcoming these barriers, moving from catastrophic frame to opportunity frame is what we are looking forward while designing climate communication strategies. The given barriers

---

[27] How Can We Make People Care About Climate Change?—Yale E360.

cannot stop motivated people to transform their behaviour and take pro-climate actions. For behavioural transformation to take place, one of the crucial enablers of action is leadership, which can inspire an innovative and collaborative organizational culture to institutionalize responses to climate change. It will also contribute to institutional adaptation, embedding SDGs in long-term strategic planning, for addressing the uncertainty associated with changing climate (Burch, 2010).

Conclusion

Climate change, a major environmental crisis of twenty first century, is influencing planet earth as a whole. Evidence shows that any change in the environment warrants changes in people's behaviour too. To draw effective solutions to this problem, there is a need to understand human behavioural responses to climate change and to switch from '*wanting change*' to actually '*working for change*'.

Transformation as an adaptive response to climate change opens a range of novel policy options (Pelling et al., 2015). Making behaviour transformation a choice to cope and adapt with climate change is also assumed to pave way for more innovative policies in this direction. The relevance of behavioural change in climate change mitigation implies that policy-informing models on climate change should include behavioural change as a complement or partial alternative to technological change (van de Ven et al., 2018). Policymaking can be a powerful way to promote behaviour transformation through both negative and positive reinforcement methods like imposing bans, introducing rules and regulations, levying taxes or even through economic incentives, subsidies, etc. However, mere policymaking is not a guarantee to ensure that people will embrace behaviour transformation for changing consumption patterns happily and willingly. Also, there are various challenges which are associated with policymaking that vary from country to country. Often there might be some political hurdles associated with the implementation of environmental policies especially in those countries where the interests of policymakers are at crossroads with the developmental interests or agenda of the governments of the concerned countries.

Global uniformity in policymaking and implementation can go a long way in bringing about the desired change in consumer behaviours with regard to sustainable consumption patterns. Policymakers may rely heavily on the field of behavioural science in improving their understanding of complex human behaviour. There will be a better scope to have meaningful impacts upon sustainable consumption choices and pro-climate action/behaviour if the field of behavioural science is integrated with policymaking. The relevance of behavioural change in climate change mitigation implies that policy-informing models on climate change should include behavioural change as a complement or partial alternative to technological change (van de Ven et al., 2018). Not only there is a need for improving upon existing sustainability policies but also there is a requirement to revamp our sustainability-related initiatives, facilities, and services. Behavioural science can give policymakers an insight into the various behavioural challenges and barriers related to consumer choices. This in turn can help fortify policies for sustainability thereby promoting more sustainable consumption behaviours. When insights from behavioural science can be transformed into real change, substantiated by contemplative practices, as being described in next chapter, we expect the outcome to be in sync with the health and well-being of the planet.

References

Adger, W. N., Barnett, J., Brown, K., Marshall, N., & O'Brien, K. (2013). Cultural dimensions of climate change impacts and adaptation. *Nature Climate Change*, 3(2), 112–117. https://doi.org/10.1038/nclimate1666

Akbar, K., Jin, Y., Mahsud, M., Akbar, M., Waheed, A., & Amin, R. (2020). Role of big five personality traits in sustainable consumption behavior. *Proceedings of the 2020 3rd International Conference on Big Data Technologies*, 222–226. https://doi.org/10.1145/3422713.3422750

Baumeister, S. (2020). Mitigating the climate change impacts of aviation through behavioural change. *Transportation Research Procedia*, 48, 2006–2017.

Blake, J. (1999). Overcoming the 'value-action gap' in environmental policy: Tensions between national policy and local experience. *Local Environment*, 4(3), 257–278. https://doi.org/10.1080/13549839908725599

Bouman, T., Verschoor, M., Albers, C. J., Böhm, G., Fisher, S. D., Poortinga, W., Whitmarsh, L., & Steg, L. (2020). When worry about climate change leads to climate action: How values, worry and personal responsibility relate to various climate actions. *Global Environmental Change*, 62, 102061.

Boyes, E., & Stanisstreet, M. (2012). environmental education for behaviour change: Which actions should be targeted? *International Journal of Science Education*, 34(10), 1591–1614. https://doi.org/10.1080/09500693.2011.584079

Brand, U. (2016). "Transformation" as a new critical orthodoxy: The strategic use of the term "Transformation" Does not prevent multiple crises. *GAIA— Ecological Perspectives for Science and Society, 25*(1), 23–27. https://doi.org/10.14512/gaia.25.1.7

Burch, S. (2010). Transforming barriers into enablers of action on climate change: Insights from three municipal case studies in British Columbia Canada. *Global Environmental Change, 20*(2), 287–297. https://doi.org/10.1016/j.gloenvcha.2009.11.009

Chiang, Y.-T., Fang, W.-T., Kaplan, U., & Ng, E. (2019). Locus of control: The mediation effect between emotional stability and pro-environmental behavior. *Sustainability, 11*(3). https://doi.org/10.3390/su11030820

Cleveland, M., & Kalamas, M. (2014) *Environmental locus of control.* https://doi.org/10.1093/acprof:oso/9780199997480.003.0009

Creutzig, F., Fernandez, B., Haberl, H., Khosla, R., Mulugetta, Y., & Seto, K. C. (2016). Beyond technology: Demand-side solutions for climate change mitigation. *Annual Review of Environment and Resources, 41,* 173–198.

Crompton, T. (2011). Finding cultural values that can transform the climate change debate. *Solutions Journal, 2*(4), 56–63.

Dietz, T., Dan, A., & Shwom, R. (2007). Support for climate change policy: Social psychological and social structural influences. *Rural Sociology, 72*(2), 185–214.

Ebi, K. L., & Semenza, J. C. (2008). Community-based adaptation to the health impacts of climate change. *American Journal of Preventive Medicine, 35*(5), 501–507.

Ersner-Hershfield, H., Wimmer, G. E., & Knutson, B. (2009). Saving for the future self: Neural measures of future self-continuity predict temporal discounting. *Social Cognitive and Affective Neuroscience, 4*(1), 85–92.

Fischer, J., Dyball, R., Fazey, I., Gross, C., Dovers, S., Ehrlich, P. R., Brulle, R. J., Christensen, C., & Borden, R. J. (2012). Human behavior and sustainability. *Frontiers in Ecology and the Environment, 10*(3), 153–160. https://doi.org/10.1890/110079

Fischer, J., Manning, A. D., Steffen, W., Rose, D. B., Daniell, K., Felton, A., Garnett, S., Gilna, B., Heinsohn, R., & Lindenmayer, D. B. (2007). Mind the sustainability gap. *Trends in Ecology & Evolution, 22*(12), 621–624.

Freud, S., & Bonaparte, P. M. (1954). *The origins of psychoanalysis* (Vol. 216). Imago London.

Giefer, M. M., Peterson, M. N., & Chen, X. (2019). Interactions among locus of control, environmental attitudes and pro-environmental behaviour in China. *Environmental Conservation, 46*(3), 234–240. https://doi.org/10.1017/S0376892919000043

Gifford, R. (2011). The dragons of inaction: Psychological barriers that limit climate change mitigation and adaptation. *American Psychologist, 66*(4), 290.

Gifford, R., Kormos, C., & McIntyre, A. (2011). Behavioral dimensions of climate change: Drivers, responses, barriers, and interventions. *Wiley Interdisciplinary Reviews: Climate Change*, 2(6), 801–827.

Gigerenzer, G., & Gaissmaier, W. (2011). Heuristic decision making. *Annual Review of Psychology*, 62, 451–482.

Haines, A., Kovats, R. S., Campbell-Lendrum, D., & Corvalán, C. (2006). Climate change and human health: Impacts, vulnerability, and mitigation. *The Lancet*, 367(9528), 2101–2109.

Hammonds, R. (2020). How can we overcome the great procrastination to respond to the climate emergency? *Health and Human Rights*, 22(1), 363.

Hardt, L., Brockway, P., Taylor, P., Barrett, J., Gross, R., & Heptonstall, P. (2019). *Modelling demand-side energy policies for climate change mitigation in the UK: A rapid evidence assessment*.

Hines, J. M., Hungerford, H. R., & Tomera, A. N. (1987). Analysis and synthesis of research on responsible environmental behavior: A meta-analysis. *The Journal of Environmental Education*, 18(2), 1–8.

IPCC. (2014). Climate Change 2014: Mitigation of climate change. In O. Edenhofer, R. Pichs-Madruga, Y. Sokona, E. Farahani, S. Kadner, K. Seyboth, A. Adler, I. Baum, S. Brunner, P. Eickemeier, B. Kriemann, J. Savolainen, S. Schlömer, C. von Stechow, T. Zwickel, & J. C. Minx (Eds.), *Contribution of working group III to the fifth assessment report of the intergovernmental panel on climate change*. Cambridge, UK and New York, NY, USA: Cambridge University Press.

Kahan, D. (2012). Why we are poles apart on climate change. *Nature News*, 488(7411), 255.

Kates, R. W., Travis, W. R., & Wilbanks, T. J. (2012). Transformational adaptation when incremental adaptations to climate change are insufficient. *Proceedings of the National Academy of Sciences*, 109(19), 7156–7161.

Keller, K., Robinson, A., Bradford, D. F., & Oppenheimer, M. (2007). The regrets of procrastination in climate policy. *Environmental Research Letters*, 2(2), 024004.

Khanna, T. M., Baiocchi, G., Callaghan, M., Creutzig, F., Guias, H., Haddaway, N. R., Hirth, L., Javaid, A., Koch, N., & Laukemper, S. (2021). A multi-country meta-analysis on the role of behavioural change in reducing energy consumption and CO2 emissions in residential buildings. *Nature Energy*, 1–8.

Kollmuss, A., & Agyeman, J. (2002). Mind the gap: Why do people act environmentally and what are the barriers to pro-environmental behavior? *Environmental Education Research*, 8(3), 239–260.

Kwasnicka, D., Dombrowski, S. U., White, M., & Sniehotta, F. (2016). Theoretical explanations for maintenance of behaviour change: A systematic review of behaviour theories. *Health Psychology Review*, 10(3), 277–296.

Lacroix, K., Gifford, R., & Chen, A. (2019). Developing and validating the Dragons of Inaction Psychological Barriers (DIPB) scale. *Journal of Environmental Psychology, 63,* 9–18.
McCarthy, J. J., Canziani, O. F., Leary, N. A., Dokken, D. J., & White, K. S. (2001). *Climate change 2001: Impacts, adaptation, and vulnerability: contribution of Working Group II to the third assessment report of the Intergovernmental Panel on Climate Change* (Vol. 2). Cambridge University Press.
Niamir, L. (2019). *Behavioural climate change mitigation: From individual energy choices to demand-side potential.*
Norgaard, K. M. (2011). *Living in denial: Climate change, emotions, and everyday life.* mit Press.
O'Brien, K., & Sygna, L. (2013). Responding to climate change: The three spheres of transformation. *Proceedings of Transformation in a Changing Climate, 16,* 23.
Pachauri, R. K., & Reisinger, A. (2007). *2007.* IPCC fourth assessment report. IPCC.
Page, N., & Page, M. (2014). Climate change: Time to do something different. *Frontiers in Psychology, 5,* 1294.
Panno, A., Cristofaro, V. D., Oliveti, C., Carrus, G., & Donati, M. A. (n.d.). Personality and environmental outcomes: The role of moral anger in channeling climate change action and pro-environmental behavior. *Analyses of Social Issues and Public Policy,* n/a(n/a). https://doi.org/10.1111/asap.12254
Park, S. E., Marshall, N. A., Jakku, E., Dowd, A. M., Howden, S. M., Mendham, E., & Fleming, A. (2012). Informing adaptation responses to climate change through theories of transformation. *Global Environmental Change, 22*(1), 115–126.
Pelling, M., O'Brien, K., & Matyas, D. (2015). Adaptation and transformation. *Climatic Change, 133*(1), 113–127.
Piligrimienė, Ž., Žukauskaitė, A., Korzilius, H., Banytė, J., & Dovalienė, A. (2020). Internal and external determinants of consumer engagement in sustainable consumption. *Sustainability, 12*(4). https://doi.org/10.3390/su12041349
Ramsey, C. E., & Rickson, R. E. (1976). Environmental knowledge and attitudes. *The Journal of Environmental Education, 8*(1), 10–18.
Ribeiro, J., Veiga, R., & Higuchi, A. (2016). Personality traits and sustainable consumption. *Revista Brasileira De Marketing, 15,* 297–313. https://doi.org/10.5585/remark.v15i3.3218
Rishi, P., & Schleyer-Lindenmann, A. (2020). Psychosocial dimensions of culture-climate connect in India and France. In W. Leal Filho, J. Luetz, &

D. Ayal (Eds.), *Handbook of climate change management: Research, leadership, transformation* (pp. 1–20). Springer International Publishing. https://doi.org/10.1007/978-3-030-22759-3_93-1

Rothermich, K., Johnson, E. K., Griffith, R. M., & Beingolea, M. M. (2021). The influence of personality traits on attitudes towards climate change— An exploratory study. *Personality and Individual Differences, 168,* 110304. https://doi.org/10.1016/j.paid.2020.110304

Roy, J., Chakravarty, D., Dasgupta, S., Chakraborty, D., Pal, S., & Ghosh, D. (2018). Where is the hope? Blending modern urban lifestyle with cultural practices in India. *Current Opinion in Environmental Sustainability, 31,* 96–103.

Ryan, R. M., & Deci, E. L. (2000). Intrinsic and extrinsic motivations: Classic definitions and new directions. *Contemporary Educational Psychology, 25*(1), 54–67.

Schlitz, M. M., Vieten, C., & Miller, E. M. (2010). Worldview transformation and the development of social consciousness. *Journal of Consciousness Studies, 17*(7–8), 18–36.

Schmuck, P., & Schultz, W. P. (2012). *Psychology of sustainable development.* Springer Science & Business Media.

Schwartz, S. H. (1977). Normative influences on altruism. In *Advances in experimental social psychology* (Vol. 10, pp. 221–279). Elsevier.

Scott, J. (2000). Rational choice theory. *Understanding Contemporary Society: Theories of the Present, 129,* 671–685.

Shiota, M. N., Papies, E. K., Preston, S. D., & Sauter, D. A. (2021). Positive affect and behavior change. *Current Opinion in Behavioral Sciences, 39,* 222–228. https://doi.org/10.1016/j.cobeha.2021.04.022

Sirois, F. M. (2014). Out of sight, out of time? A meta-analytic investigation of procrastination and time perspective. *European Journal of Personality, 28*(5), 511–520.

Soutter, A. R. B., Bates, T. C., & Mõttus, R. (2020). Big five and HEXACO personality traits, proenvironmental attitudes, and behaviors: A meta-analysis. *Perspectives on Psychological Science, 15*(4), 913–941. https://doi.org/10.1177/1745691620903019

Stern, P. C., Dietz, T., Abel, T., Guagnano, G. A., & Kalof, L. (1999). A value-belief-norm theory of support for social movements: The case of environmentalism. *Human Ecology Review,* 81–97.

Stoknes, P. E. (2014). Rethinking climate communications and the "psychological climate paradox." *Energy Research & Social Science, 1,* 161–170.

Stoknes, P. E. (2015). *What we think about when we try not to think about global warming: Toward a new psychology of climate action.* Chelsea Green Publishing.

Sussman, R., Gifford, R., & Abrahamse, W. (2016). *Social mobilization: How to encourage action on climate change*. Pacific Institute for Climate Solutions.

Swim, J. K., Stern, P. C., Doherty, T. J., Clayton, S., Reser, J. P., Weber, E. U., Gifford, R., & Howard, G. S. (2011). Psychology's contributions to understanding and addressing global climate change. *American Psychologist, 66*(4), 241.

Tompkins, E. L., & Adger, W. N. (2004). Does adaptive management of natural resources enhance resilience to climate change? *Ecology and Society, 9*(2).

van de Ven, D.-J., González-Eguino, M., & Arto, I. (2018). The potential of behavioural change for climate change mitigation: A case study for the European Union. *Mitigation and Adaptation Strategies for Global Change, 23*(6), 853–886.

Verplanken, B. (2018). Promoting sustainability: Towards a segmentation model of individual and household behaviour and behaviour change. *Sustainable Development, 26*(3), 193–205.

Wang, P., Liu, Q., & Qi, Y. (2014). Factors influencing sustainable consumption behaviors: A survey of the rural residents in China. *Journal of Cleaner Production, 63*, 152–165.

White, K., Habib, R., & Hardisty, D. J. (2019). How to SHIFT consumer behaviors to be more sustainable: A literature review and guiding framework. *Journal of Marketing, 83*(3), 22–49.

Wi, A., & Chang, C.-H. (2019). Promoting pro-environmental behaviour in a community in Singapore–From raising awareness to behavioural change. *Environmental Education Research, 25*(7), 1019–1037.

Williamson, K., Satre-Meloy, A., Velasco, K., & Green, K. (2018). Climate change needs behavior change: Making the case for behavioral solutions to reduce global warming. *Rare: Arlington, VA, USA*.

Zhao, C., Yan, Y., Wang, C., Tang, M., Wu, G., Ding, D., & Song, Y. (2018). Adaptation and mitigation for combating climate change–from single to joint. *Ecosystem Health and Sustainability, 4*(4), 85–94.

CHAPTER 7

Contemplative Practices, Climate Change Adaptation, and Sustainability Management

INTRODUCTION

Global society is under threat in view of systematic and intensified dilapidation of socio-ecological systems at a massive scale and humanity's role in changing the climate is considered as an assault on the divine natural creation (Broman & Robèrt, 2017). We, as human beings, with faith in any religion, are concerned about this ongoing destruction of natural world and worry about the future of planet earth. An array of challenges linked to sustainability and climate change, being faced by humanity cannot be exclusively addressed through policy, governance, and technological interventions (Abson et al., 2017; Wamsler, 2019). Gus Speth, a US advisor on climate change rightly remarks '*I used to think the top environmental problems were biodiversity loss, ecosystem collapse and climate change. I thought with 30 years of good science we could address those problems. But I was wrong. The top environmental problems are selfishness, greed and apathy... And to deal with these we need a spiritual and cultural transformation - and we scientists don't know how to do that*'.[1]

Robèrt et al. (2013) provided a methodical approach to sustainability in a Framework for Strategic Sustainable Development (FSSD), explaining

---

[1] http://winewaterwatch.org/2016/05/we-scientists-dont-know-how-to-do-that-what-a-commentary/.

© The Author(s), under exclusive license to Springer Nature Singapore Pte Ltd. 2022
P. Rishi, *Managing Climate Change and Sustainability through Behavioural Transformation*, Sustainable Development Goals Series, https://doi.org/10.1007/978-981-16-8519-4_7

it as a '*strategic mission to eliminate society's unsustainable structures and behaviours, and create a sustainable society operating within the carrying capacity of socio-ecological systems*'. Therefore, for the sake of prosperity of human life on planet earth, it is imperative that human wants and needs are fulfilled taking into consideration the carrying capacity of our planet.

Climate change is giving rise to perceptible sustainability challenges, requiring need for innovative pathways to promote climate adaptation (Sol & Wals, 2015). In this regard, alignment of behavioural science and philosophical concerns with sustainability and climate change can give good results in addressing the emerging challenges (Broman & Robèrt, 2017). In the recent past, there has been a consequential shift in climate change and sustainability research embracing the psychological, philosophical, and socio-cultural dimensions with a focus on attitudes, motivations, beliefs, and values. Hence, climate change and sustainability are also considered as moral and spiritual issues and contemplative practices have become important to be empirically explored in this context from transdisciplinary perspective. The statement by Al Gore while receiving the Nobel prize is worth mentioning in this context—'*We face a true planetary emergency. The climate crisis is not a political issue, it is a moral and spiritual challenge to all of humanity. It is also our greatest opportunity to lift global consciousness to a higher level*'.[2]

The chapter relates Climate Change (CC) linked sustainability to psycho-spiritual reality from a universal belief-based perspective with the aim to strengthen behavioural adaptation to changing climatic scenario. The broad thematic structure of the chapter starts with philosophical and psycho-spiritual basis of sustainability, taking into consideration the inner world and role of contemplative practices like mindfulness and spiritual intelligence to establish the human–sustainability connect which can possibly result in behavioural transformation for sustainability and climate change adaptation. It is explained through-

a. Enhancing the knowledge base in psycho-spiritual systems.
b. Ways to overcome the psychological barriers to climate change adaptation and sustainability management.

---

[2] https://www.reuters.com/article/us-nobel-peace-gore-reaction-idUSN1245757720 071012.

c. Developing the capacity to address climate change linked adversities through mindfulness and other socio-behavioural techniques.

Further, application of psycho-spiritual concepts will be explained through Positive Psychology Intervention Programme (PPI) along with their implications to benefit the environment and sustainable living. Particularly, it will focus on Application of Mindfulness (MI) and Spiritual Intelligence (SI), Behavioural Restraint (BR), and Positive Existential Transcendence (PET) and establish the need for cross-country collaborative research on the role of different contemplative practices in the east and the west.

Perspectives of Climate Change and Sustainability

Climate change and sustainability are explained from the perspectives of both natural sciences as well as human sciences. While natural sciences explain the dos and don'ts of coping with adversities of climate change and sustainability crisis and suggests creating structural options for preparedness and mitigation, human sciences focus on associated personal, social, and spiritual belief systems, perceiving climate risk and motivation to practice established ways of coping with climatic stressors. Climate change and sustainability management is interpreted from multiple perspectives as depicted in Fig. 7.1.

The climate and sustainability sciences start with the biophysical perspective which used to be considered as the mainstream approach till subsequent perspectives like socio-cultural and psychological gradually gained acceptance. Further, behavioural scientists and philosophers started integrating philosophical and spiritual dimensions to promote sustainability behavioural transformation.

The term sustainability is explained from traditional as well as modern viewpoints as depicted in Fig. 7.2. Considering the importance of modern viewpoint of sustainability and SDG 3. It emphasizes well-being, from a psychological perspective, promoting enrichment, growth, and flexible change. The importance of psycho-spiritual paradigm in interpreting climate change adaptation and sustainability management is further enhanced in this context.

Sustainability action requires the triad of systems' knowledge comprising of technical know-how, normative knowledge comprising of

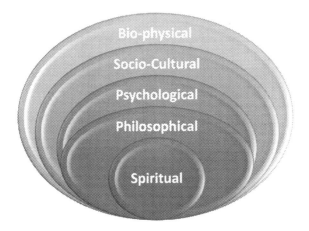

Fig. 7.1 Dimensions of global climate change and sustainability management

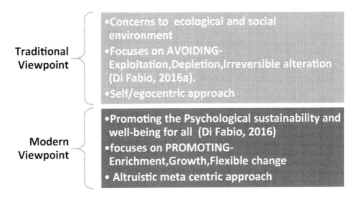

Fig. 7.2 Viewpoints of sustainability

'socio-cultural know-how' and transformative knowledge comprising of 'individual mindset know-how' (Abson et al., 2014).

Perspectives of climate change (CC) start with risk perception and associated distress as a part of identification of climate risk which leads to how people attribute reasons to climate change and engage in mitigation and adaptation strategies. It has already been described in detail in chapter two. This whole process of climate risk identification and individual/societal choice of attribution/mitigation/adaptation is possibly

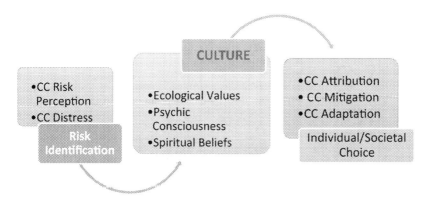

Fig. 7.3 Dynamics of climate change—psycho-spiritual dimensions

mediated by ecological values and psychic consciousness coupled with spiritual belief systems as reflected in Fig. 7.3.

Hence, socio-cultural, psychological, and spiritual factors play an equally important role in determining the way people adapt with climate change at the behavioural level. However, intensive empirical research is needed to substantiate this conceptual framework in diverse geographical regions in view of the cultural and context specific attributes of climate change adaptive mechanisms.

Psycho-Spiritual Basis of Sustainability

It is evident that in spite of our long cherishing efforts to promote environmental sustainability and combat climate change at multiple fronts, integrating best possible technology-centric mitigation and adaptive mechanisms, the degradation of the ecosystem is still continued and the socio-economic inequality is still rising, posing serious questions to the expertise and adequacy of sustainability scientists (Ives et al., 2020). WWF (2016) remarks that 'despite the prominence of sustainability as a concept, planetary trajectories remain deeply unsustainable'.[3] So the question arises, that in the facade of fascinating climate change mitigating/adaptive strategies, are we not losing out on the human genre

---

[3] link.springer.com/article/10.1007/s13280-019-01187-w.

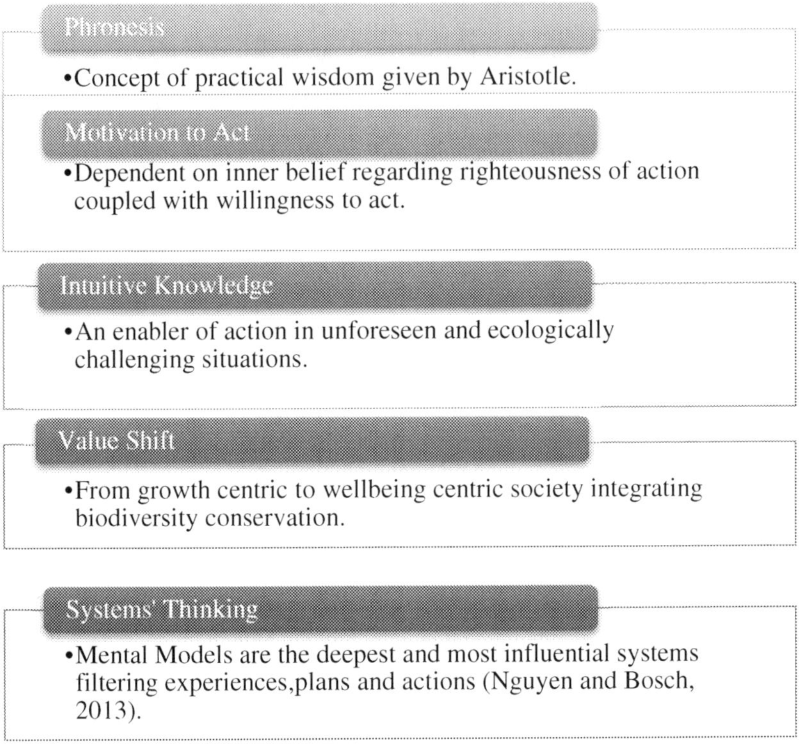

Phronesis
- Concept of practical wisdom given by Aristotle.

Motivation to Act
- Dependent on inner belief regarding righteousness of action coupled with willingness to act.

Intuitive Knowledge
- An enabler of action in unforeseen and ecologically challenging situations.

Value Shift
- From growth centric to wellbeing centric society integrating biodiversity conservation.

Systems' Thinking
- Mental Models are the deepest and most influential systems filtering experiences, plans and actions (Nguyen and Bosch, 2013).

Fig. 7.4 Progression of psycho-philosophical thoughts of sustainability

while assessing their efficacy in practical terms? Are we probably preoccupied to accept everything, what is scientifically proven, as acceptable to people and society, no matter whether they fit in their individual mindsets, collective social structures, and cultural milieu or not? To explore it further, we need to explore the progression of psycho-philosophical thoughts on sustainability as reflected in Fig. 7.4.

Many people are cognitively aware of the knowledge of conservation of resources and pro-climate action but in the context of sustainability, *Phronesis*[4] is the practical wisdom which supports to convert this instinctive pro-climate/sustainability knowledge into practice. The

[4] https://phronesis.pressbooks.com/chapter/virtue-ethics/.

concept given by Aristotle has been well explained by Schwartz and Sharpe (2006) while relating it to positive psychology. Subsequently, even if there is a presence of practical wisdom, it requires '*motivation to act*' which is dependent on how strongly one believes to execute pro-climate/sustainability action at affective level along with willingness to act (Hume, 1975). Further to that, the concept of '*Intuitive Knowledge*' which enables a person to act in ecologically challenging situations (Harding, 2006). Next in the line is the thought process of '*value shift*' where need to move from 'growth centric' to 'wellbeing centric' society integrating biodiversity conservation was expressed (Manfredo et al., 2017; Martin et al., 2016). Recent trends in psycho ecological thoughts progressed to '*system thinking*' with a focus on mental models as the systems to filter the conservation-centric experiences, plans, and actions (Nguyen & Bosch, 2013).

Further, Wamsler (2017) used the term 'inner transition' to explain the psychological shift within human beings that connects to their 'expanded consciousness' leading to values-led behavioural transformation. In the same context, Ives et al. (2020) also try to establish the importance of inner world comprising of feelings, emotions, beliefs, thoughts, and social identities, as the heart and soul for sustainability, which has the power to transform systemic changes. Lama (2001) in his book, 'Ethics for the New Millennium', states that thoughtfulness to our inner worlds would both lead to greater happiness and well-being besides laying a sound underpinning for an ethical and sustainable global society. Individual characteristics like compassionate and empathic behaviour towards environment also gives expressions to ecologically sustainable practices, helping in addressing the issues of climate change directly or indirectly.

However, these inner realities, being dependent on subjective, experiential, and intuitive modes of understanding, are often neglected by mainstream sustainability scientists, resulting in sustainability crisis.

Sustainability crisis is an outcome of our untapped inner realities which can only be managed through contemplative practices, more commonly practised in the eastern world. Psycho-spirituality contains the underpinning assumption that the psychological mindset creates, or strongly influences spirituality and suggests that one can learn to adapt with the changing climatic scenario through spiritual modes. It can subconsciously control materialistic urges leading to ecologically unsustainable growth and development on one hand and development of positive motivations and emotions towards changing climatic scenario on the other. Inner realities like compassion, empathy, emotions (Wamsler, 2019) coupled with contemplative practices like mindfulness, behavioural restraint, and positive existential transcendence can help to adapt with climate change and manage the sustainability of planet earth.

Role of Contemplative Practices

Natural environments have the property of resilience to bounce back in order to survive the unsettling impacts without any major physical/psychic deterioration (Martin-Breen & Anderies, 2011). Similarly, human beings also have the auto-repair ability to safeguard themselves from a variety of threats constituting human resilience. Contemplative practices dynamically act to channelize tranquillity and absorbed attention to assure internal stability. It is attained by fostering a greater sense of peace and well-being, extenuating tension, and honing attention and concentration skills. Further, these practices also expand connections with nature and the natural world by increasing our ability to 'challenge, investigate and engage with rapid change and uncertainty' (Allison, 2016).[5]

The role of contemplative activities in fostering resilience at group and community level is identified in research work that could promote resilience in five human domains including psycho-behavioural domains, brain function, stress physiology, healthy gene expression, and post-traumatic growth, which results into increased attention and emotional regulation, develop neural plasticity, optimize stress response, and improve psycho-social functioning through meditation and yoga practices (Sivilli & Pace, 2014). Contemplative practices such as yoga, meditation, reflection, and breathing have also been explored against resilience in the sustainability classroom, revealing that these practices lead to an increase in learner resilience and the ability of learners to indulge in complicated socio-ecological resilience (Goralnik & Marcus, 2020). Meditation, as a contemplative exercise is routinely performed by Buddhist monks. While studying the brain scans of these monks performing compassion meditation, a rise in high-frequency gamma waves has been recorded that

---

[5] Allison, E. 2016. Cultivating resilience for addressing ecological change. Embrace of the Earth 2016. 8. [online] URL: https://digitalcommons.ciis.edu/embraceoftheearth2016/8.

underlie elevated mental activity such as consciousness (Duerr, 2011). Loving kindness meditation practice and interventions are found to be effective in enhancing positive emotions (Zeng et al., 2015).

Issues of sustainability have deep rooted association with the inner world of human beings. Contemplative practices can be instrumental in establishing the connect with inner world through development of vital cognitive skills (Papenfuss, 2019; Papenfuss et al., 2019) which can help in promoting sustainable behavioural transformation of individuals and society along with addressing the sustainability challenges. Some of the important contemplative practices which have the direct role in addressing the issues of climate change and sustainability, as proclaimed in research include spiritual intelligence and mindfulness as described below.

Spiritual Intelligence (SI)

The rich cultural legacy of the east is preserved and nurtured by philosophical values and traditions within the sense of spirituality. Spiritual belief systems have tended to conserve plant and animal populations in the hands of hunter gatherer groups with poor literacy rates and limited scientific temperament, especially in India. Moreover, India's traditional understanding of biodiversity conservation is diverse with 2753 population groups and their geographic distribution, agriculture techniques, eating preferences, lifestyle techniques, and cultural practices (Singh et al., 2014).

Concept of Spirituality and Spiritual Intelligence
Spirituality comprises of four proficiencies, namely, the experience of an elevated level of self-consciousness; application of spiritual resources to resolve life challenges and good acts such as redemption, self-acceptance; the investigation and filtration of everyday experiences and feelings; and the desire to subsume acts in order to facilitate the development of the planet that corresponds to greater environment adaptability in individuals (Ajawani, 2017). The spiritual intelligence framework integrates science and spirituality as it is a direct experience and intrinsic capability, broadly unrelated to religious belief system. Hence, it is legitimized in contemporary secular world, free from religion-centric biases. (Emmons, 2000) explored spirituality as a form of intelligence allowing people to resolve problems of everyday lives and attaining several goals through five components such as transcendence capacity; heightened consciousness state;

sense of sacred in daily activities; events and relationships; engagement in virtual behaviour. Amram (2007) narrated seven dimensions of spiritual intelligence for adaptive application in daily work activities that include consciousness, meaning, grace, transcendence, truth, peaceful surrender to self, and inner directedness across different spiritual traditions.

Spiritual Intelligence is described as the ultimate intelligence which is positioned at the top of the hierarchy, in contrast to Emotional Intelligence and Rational Intelligence by (Zohar & Marshall, 2001). It is also defined as *'The ability to act with Wisdom and Compassion while maintaining inner and outer peace (equanimity), regardless of the circumstances'*.[6] It is reflected through wisdom, compassion, integrity, love and joy, peace and creativity, the attributes, which have special relevance for promoting pro-climate and sustainability behaviour (Fig. 7.5). Spiritual Intelligence exists as a potential in each of us—but like any intelligence it must be developed. Developed intelligences can be demonstrated as 'skills' or 'competencies'. It enables the optimal functioning of both the Intelligent Quotient (IQ) and the Emotional Intelligence by magnifying and incorporating them (Emmons, 2000).

Neural Basis of Spiritual Intelligence
According to Richard Griffiths,[7] spirituality is having a strong neutral basis. The neutral model of intelligence speaks about different forms of information processing linked to different parts of the brain. Cognitive intelligence is the prerequisite of left brain which involves serial processing while emotional intelligence pertains to right brain involving parallel processing of information. However, spiritual intelligence involves whole brain synchronizing the information processing between left and the right brain. Hence, it is considered as highest level of intelligence and can be applied in diverse fields to derive positive outcomes.

Neurological research proclaim that the experience of presence is closely connected to hemispheric synchronization and activation of the whole brain thus connecting mind, self, and external world together into a holistic and meaningful sphere. Studies prove that spiritual intelligence is hard-wired in the human brain but needs intentional activation to derive its benefits. Hemispheric synchronization activates whole brain

---

[6] https://olstraining.com/sq21-sprutual-intelligence-assessment-w-o-debrief/.
[7] https://sqi.co/practitioner-training-course-spiritual-intelligence.

Fig. 7.5 Reflections of spiritual intelligence[8]

achieve coherence and optimizes the functionality of human brain function resulting in sense of fulfilment, heightened creativity, intuition, empathy, and compassion, and resulting positive outcomes on diverse academic, social, and other activities. Gamma frequencies are also associated with SI activation which otherwise occur rarely and at random. SI also stimulates specific regions of the brain, accountable for neurogenesis, the process to strengthen existing and building the new brain networks. Neurotransmitters generating fight and flight reactions are also reduced and those associated with well-being and balance are increased by well-developed SI.

The above analysis of neurological basis of SI potentiates its pertinence to be applied for promoting climate action and sustainability behaviour.

[8] https://sqi.co/practitioner-training-course-spiritual-intelligence.

Spiritual Intelligence for Climate Action and Sustainability Behaviour

Spiritual intelligence and belief systems are actively instrumental in framing the psychological subset of attitudes across cultures, which subsequently determine the way climate action and sustainability behaviour of people takes place. Spiritual intelligence allows incorporating consciousness of inner life and energy into the outside life of function in the environment and can be achieved by practice, inquiry, and questing (Vaughan, 2002). The idea of spiritual intelligence is described by (Zohar, 2012) as doing things in a right way through sense of knowledge and awareness of the inner self. Hence, it can be used as a tool for appraising climate risk and ascertaining adaptive capacity of people regarding climate change (Fig. 7.6).

Contemporary spirituality can be explained in context to its potential and pitfalls for sustainable development. According to (Hedlund-de Witt, 2011), contemporary spirituality requires the ultimate restoration of nature, which is reflected in relation to organic food and vegetarian diets, recycled foods, and mindful consumerism. It results in less contamination of the environment and attenuation of resources by greening of human lifestyles, which also facilitates and encourages the development of a green

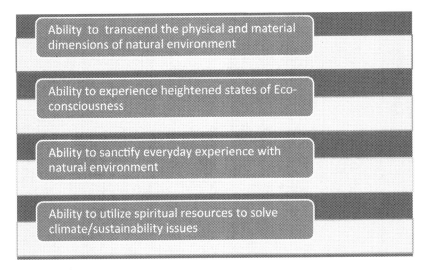

Fig. 7.6 Spiritual intelligence for climate change and sustainability

economy. While appraising the connection between resilience and spiritual intelligence (SI), a positive relation was found among them, suggesting that SI helps to overcome negative effects of stress by strengthening resilience (Khosravi & Nikmanesh, 2014). In regard to the psychological component of SI, it helps to accomplish the potentials of human abilities by non-cognitive attributes in order to deal with daily work functions of life, creatively and constructively in the socio-psycho-physical setting leading to the highest degree of awareness and knowledge (Srivastava, 2016).

In the context of climate action and sustainability behaviour, spiritual intelligence can facilitate creative behavioural transformation over and above the cognitive attributes of sustainability, suiting to socio-cultural context to attain environmental desirable outcomes. Various aspects of spirituality, such as religious attitudes, ethical sensitivity, and harmony, when linked to adoption of coping mechanisms with environmental stress, contribute to a directly proportional association between the two. When undergoing climatically distressing situation, individuals with high spirituality would seek a solution through task-oriented and social diversion coping styles (Krok, 2008).

Finally, we can say that people are increasingly in urgent need of behavioural transformation to address the challenges of sustainability and to cope with the various adverse consequences of climate change. Integrating spiritual knowledge in day-to-day behavioural practices can play a positive role by facilitating subjective well-being and enhancing the climate resilience on one hand and promoting behavioural transformation as well as adaptation on the other. Dedicated empirical research in diverse socio-cultural contexts can further substantiate these assertions.

Mindfulness

The empirical data base on climate change and sustainability makes it clear, that the issue cannot simply be resolved by adoption of technology or government/ policy interventions. There is a need to develop behavioural/social practices and embolden a far-reaching cultural shift towards more sustainable living and climate action. It requires a complete rethinking of how we interface with our natural environment and this is where mindfulness comes in.

Mindfulness, though alternatively reflected in research literature as a state, trait, or training (Grossenbacher & Quaglia, 2017), is central to the upcoming global interest in contemplative climate change and sustainability research. Originated from the eastern Buddhist traditions, characterizing distinctive psychological processing contributing to holistic and caring life (Dreyfus, 2011), mindfulness is the process of intentionally attending moment by moment with openness while being non-judgmental. Subsequently, another measure explains mindfulness as sustained present-moment awareness (Levinson et al., 2014).

Principle of Dependent Origination
The fundamental base of the principle of dependent origination, originated in the ancient Buddhist literature is that nature or life is resting on a set of relations, in which rising, or cessation of elements depend on the dynamics of factors conditioning them. *'We learn from the principle of dependent origination that things and events do not come into being without causes. Suffering and unsatisfactory conditions are caused by our own delusions and the contaminated actions induced by them'.* remarks, 14th Dalai Lama.

It is also interpreted as the principle of interdependence and relativity, which means that the arising, continuation, and cessation of existence of nature is dependent on the web of interrelated factors as all universal phenomena are relative to each other. They cannot exist independently without the combination of supportive conditions and will cease when the supportive conditions causing change are no longer able to sustain it. Hence, supportive conditions play a decisive role in arising, sustenance, and cessation of elements of nature.

Applying this principle to human–environment relationship is in consonance with the scientific approach. The survival and sustenance of this phenomenal natural world is dependent on the way it relates to the heart and soul of people who are its so-called custodians (or destroyers). Humans have a fictitious belief about permanence of natural environment and its gifts to humankind irrespective of how they treat them leading to dominance of forces of greed. This illusion of the permanency of enriched natural environment makes us difficult to believe that the natural world is like a bubble and not the kind of reality which is bound to burst if we are not able to handle it with compassion and care.

The fundamental principle of dependent origination talks about cause and effect. To illustrate further, relating to the things around us, let us take the flame of an oil lamp which is dependent upon the presence of oil and the wick and in absence will cease burning. Similarly, germination of a sprout is dependent upon the seed, moisture, soil, air, and sunlight and cannot function independently. Similar is the scenario with flowing of river which is dependent on the continuation of stream one after the other. The ignorance of human beings to understand this relationship and interdependence with natural environment can lead to irresponsible behaviour causing climate distress or sustainability crisis while the understanding can enhance environmental consciousness and care and compassion towards processes of nature.[9]

Taken together, all human beings are deeply connected to other beings and the natural world, including their actions, and thinking. One environmentally irresponsible action by one person will have a chain impact on everyone on this planet earth. Hence, promoting mindful behaviour towards nature through compassion and care is extremely important to ensure the survival and sustenance of mother earth.

Mindful-Climate Adaptation and Action
When mindfulness is associated with ecologically responsible behaviour that is oriented to the common good. It promotes the integration and blending of thought towards sustainability. Behavioural adaptation with climate change is intentionally creating changes in existing behaviours and lifestyle patterns to adjust with occurring or likely to occur climatic changes in order to save earth from further degradation and save humans from deteriorated quality of life and climatic distress. Mindfulness has the property to potentiate intentional changes in behaviour. It has a positive influence on subjective well-being, activation of non-materialistic core values, consumption and sustainable behaviour, equity issues, socio-political structures, and broadly, human–nature connection. American Psychological Association has viewed mindfulness as a strategy to strengthen relations with the natural environment and to foster sustainable attitudes (Liu & Valente, 2018). Carroll (2016) considered mindfulness as a prerequisite to compassionately connect with 'socio-ecological crises'[10] (Fig. 7.7).

---

[9] https://www.budsas.org/ebud/whatbudbeliev/106.htm.

[10] Carroll, M. (2016). Formless meditation and sustainability. In *Spirituality and Sustainability* (pp. 139–153). Springer, Cham.

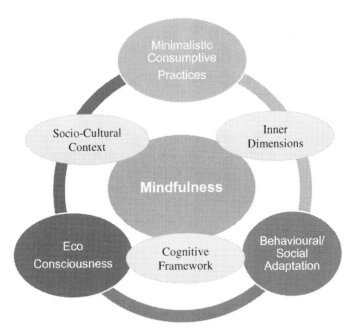

Fig. 7.7 Framework of mindfulness for climate change/sustainability

Researchers are, nowadays, focusing on the interaction between inner aspects of human beings and associated changes as a response for ever-increasing climate change that could not be overcome by current modern technologies and policies alone. The above framework suggests that mindfulness is at the centre-stage to promote eco-consciousness, minimalistic consumptive practices, and behavioural/social adaptation with climate change/sustainability crisis. However, this impact/influence is likely to be mediated by inner dimensions, cognitive framework, and socio-cultural context. Further empirical research in diverse settings may be required to potentiate this framework.

While highlighting the importance of inner transition (Wamsler et al., 2018) states that it is possible primarily through indigenous spiritual practice originated in the east called mindfulness. Mindfulness can lead to new paths for sustainability. Individual's inner mindfulness is correlated with climate adaptation at various scales, which has contributed to a higher motivation for taking climate adaptation action; identification of climate change and its perception of risk; and the withdrawal of people from fatalistic attitudes (Wamsler & Brink, 2018). Mindfulness has the potential to facilitate adaptation at all levels (person, institutional, social, and research) through perceptual, institutional, structural, ontological, and epistemological transition processes and should therefore become a central factor in climate and related resilience science. He further talks about 'mindful climate adaptation', and the need for empirical studies to explore the role of individual mindfulness in climate adaptation.

A novel approach of 'contemplative cognition framework', comprising of attentional processes—'Intended Attention, Attention to Intent and Awareness of Transient Information', overcomes the gaps in mindfulness studies by encouraging motivational and theoretical dimensions of research and facilitating studies from phenomenological, cognitive processing, neurophysiological, and clinical perspectives (Grossenbacher & Quaglia, 2017) (Fig. 7.8).

High rate of resource consumption in population intensive eastern world as well as carbon intensive western world, is further escalating the adverse impacts of climate change at a faster pace reflecting the need for constraining resource use and subjective well-being. That is primarily the idea behind sustainability. If subjective well-being is accomplished with minimized material consumption, sustainable behaviour can be seen.

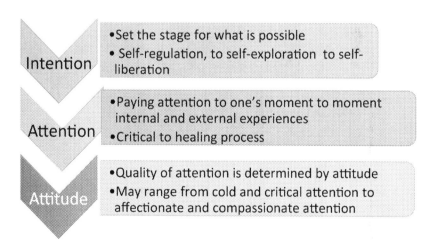

Fig. 7.8 Intention-attention-attitude framework of mindfulness

Contemplative practice such as mindfulness leads to subjective well-being resulting in improved empathy, compassion, clarity of goals and values, and both of these aspects are correlated with more sustainable behaviour, as shown in the studies (Ericson et al., 2014).

Mindfulness is associated with readiness to change and is explored at the individual and organizational level. At the individual level, it illustrates cognitive processes by improving willingness via making employee behaviours and perceived understanding more adaptive and by enhancing perceived control and self-efficiency (Gärtner, 2013). Amel et al. (2009) observed that mindfulness in all its aspects is worth promoting as an external indicator of positive environmental actions and to reach ecological resilience, we will need a mix of greater understanding and more widely accessible and viable alternatives.

Mindfulness for Sustainability Behaviour
The idea of sustainability is increasingly attracting popularity due to its relevance to the well-being of organizations at all levels. The psychology of sustainability and sustainable growth is associated with raising the quality of life of every human being, along with improvements in the natural, economic, and social system (Di Fabio, 2017). Sustainable capability development approach is preferred for solving challenges of the present time which combines philosophy by putting humanity at first place in the ecosphere for building mindful capabilities, which results from contemplative practices like meditation (Mabsout, 2015).

Mindfulness, with a focus on sustainability, supports the understanding that mindfulness can be key to politically sensitizing people and organizations to the consequences of unquestioned structures, power relations, and consumer behaviours. It helps people feel more closely connected to and understand the impacts of their own behaviour on distant communities and on the environment as a whole.

Mindfulness enhances the awareness of self-nature interconnectedness and how our actions can have a direct effect on the environment and ecosystem. Through tools of measuring mindfulness, we can observe the shifts in mindful thinking in sustainability context after providing mindfulness training. Two of the facets of Five Factor Mindfulness Questionnaire are correlated with a higher frequency of engagement in pro-environmental behaviours like noticing visual elements and in difficult situations, taking pause without immediately reacting (Barbaro & Pickett, 2016). Mindfulness intensifies experiences with the natural environment, which may foster a stronger connection with the natural world, and in turn, may regulate behaviour by making sustainable options more salient. Recent advances in neuroscience and other fields suggest that mindfulness can open new pathways towards sustainability. Neuroscience and neuroplasticity linked scientific and popular literature also suggests that mindfulness can rewire our brains (Doty, 2016) and may be a necessary component of the conversion to a more sustainable society (Koger, 2015). Based on a survey of citizens at risk of severe climate events, it was found that individual mindfulness may be able to facilitate sustainable climate adaptation (Wamsler & Brink, 2018).

Mindful individuals perceive stress in a particular manner and use a more approach-oriented coping strategy that results directly in resilience (Sivilli & Pace, 2014). If we interpret this research from environmental perspective, we can deduce that mindful individuals are more likely to sustain and cope with various environmental and climate stressors and can exhibit relatively better climate resilience, though empirical research in various geographical context is needed to substantiate this proposition. Brown and Ryan (2003), empirically emphasized that individuals who do not practice mindfulness frequently, end up in unsustainable activities in daily life, while, on the other hand, mindfulness allows people to stay away from unconscious thoughts and to become more open to behavioural improvement and ability to make alternative decisions in favour of sustainability.

Mindless consumption is passive consumption that includes actions like buying something that was not a planned purchase because it is on sale. Overconsumption correlates to less happiness, less financial wealth, lower self-esteem, more anxiety, and poorer social relationships (Chancellor & Lyubomirsky, 2011). Material consumption may stem from a feeling of impulsiveness, a psychological need for status, and boredom. It can become a form of self-medication to soothe these feelings. Advertisements also play a critical role in materialism and mindless buying because they tell consumers that buying more means living a happier life, which is not actually true. Mindfulness improves subjective well-being, which is linked to higher self-esteem and greater satisfaction with life. By feeling content with oneself without seeking approval from others, a mindful person satisfies psychological needs through spiritual experiences, is less susceptible to marketing tactics, and does not consume to find fulfillment.

Higher mindfulness means, higher overall motivation for taking (or supporting) climate adaptation actions, being motivated by social factors to take adaptation action (i.e. being encouraged by friends and family, reducing the risk for others, and having a 'good conscience'), having taken social ('other-focused') adaptation measures that require community interaction (here, warning people about environmentally hazardous events); belief in (correct representation of) climate change; pro-environmental behaviour (like following vegetarian dietary patterns); non-fatalist attitudes and many other similar instances. So, it is no longer just a concept to address cognitions, but also a tool to foster a sense of person–environment positive connection.

Finally, it can be inferred that mindfulness has a broad potential to solve the numerous problems that our civilization is facing. If we study the role of mindfulness in the light of climate change, we will note that this element of mindfulness is very recent in the world and that scholars are able to investigate this feature to a restricted degree. Concepts such as mindful-climate adaptation, relation with resilience and sustainability have been introduced but limited empirical work on the climate change dimension have been reported in both eastern as well as western countries. Hence more research is needed to assess the value of this contemplative practice for the development of sustainable and environment friendly society through effective and long-term behavioural transformation.

Taking into account the function of mindfulness in sustainability research, practice, and training (Wamsler et al., 2018), it has been documented that it has a significant effect on psychological well-being, the activation of core beliefs, utilization and sustainable actions, human–nature linkages, equity concerns, social advocacy, committed or responsive or adaptive responses. This will lead to a greater understanding of sustainability at all levels including individual level through promotion of behavioural restraint and should therefore be of utmost significance in sustainability research, practice, and teaching.

Behavioural Restraint for Sustainability Behaviour
Materialism promotes environmental, socio-psychological ills like adverse impacts of climate change, degradation, and depletion of natural resource. It also leads to possessiveness, dependence, and psychic attachment with material world. Behavioural Restraint is the ability to exert intentional and voluntary control to organize the behaviour in a manner that it is not causing any adverse impact on the environment and sustainability. Despite enjoying the world of pleasures, luxuries, and consumables, people with behavioural restraint are not mentally attached to them rather they are

being used mindfully as the modes of serving basic requirements sparingly as and when required and are not being used as the status symbols. In this manner they are able to draw continuous happiness and satisfaction by following the simplistic model of life which is a blessing in disguise with its invisible virtues. Behavioural restraint counters the psychological distress arising out of complexities of life, rising temptations, and unrealistic desires. It is possible only if the world can proceed for materialism to supra-materialism and ultimately achieve the balance between the two (Rishi, 2009).

Mindfulness can possibly be one of the tools to exercise behavioural restraint for sustainability behaviour. The eastern spiritual text *'Bhagwat Gita'* also narrates the importance of behavioural restraint (also known as supra-materialism) for eternal happiness. This state of equilibrium is also known as 'Yoga' or the state of restraint from worldly desires and control over mind and body. Control over senses stabilizes the mind while materialism excites sense organs. Total restraint from materialism is also practically difficult for people with limited spiritual inclination and, at the same time, detrimental to growth and development of a society. However, over-indulgence in materialism and consumerism also leaves its long-lasting adverse effects on the psyche as well as sustainable existence of planet earth. Therefore, the line of balance between the two deriving the essence of the eastern contemplation with the western energy intensive developmental paradigm is to be established for the creation and sustenance of sustainable world.

Contemplative practices are based on compassion and positive emotion, unlike crisis approach or motivation by fear, which are often used in climate change communications. Positive and sustainable lifestyle changes stem from increased positive feelings and a connection to the natural world. In this regard, Positive Psychology Interventions (PPI) can play an important role to promote sustainable behaviour in a non-threatening manner.

Positive Psychology Interventions (PPI) for Sustainability Behaviour
Sustainability behaviour attempts to conserve the socio-physical environment for the present as well as future generation. The marriage of human and social sciences integrating behavioural science can serve this objective. Psychological attributes and contextual factors are primarily accountable for adoption of sustainable lifestyles which can be empirically and meaningfully explored by behavioural sciences to explore how

sustainable behaviours can promote climate resilience as well as uphold subjective well-being, thus, moving towards the UN SDG 3 (Health and Well-being) and SDG 13 (Climate Action).

The genesis and answers to emerging sustainability crisis can be explained through environmental psychology paradigms suggesting that human cognitions and emotions can be modulated to address the threats to ecological issues like climate change, increasing pollutants in environment, biodiversity loss, and diminishing resources (Verdugo, 2012). The study further suggests the linkages of these ecological issues to some of the societal challenges like weather stressors and financial inequalities (heat, cold, famine, etc.). Psychological predispositions like cognitions, motives, emotions, altruistic tendencies, and behavioural capacities have the potential to promote sustainable lifestyles. Thus, psychological perspectives of environmental stressors can offer behavioural solutions to sustainability crisis and can effectively promote human well-being.

Promoting sustainability behaviour can be negatively guided in the form of instigating fear or guilt in the minds of people involved in creating sustainability crisis through environmentally irresponsible behaviour resulting in adverse behavioural outcomes like discomfort and inconvenience. Alternatively, positive psychology proclaims building on human capacities, positive emotions, and environment friendly behaviour resulting in positive behavioural outcomes like satisfaction and well-being, as noteworthy factors behind climate and environment friendly actions (Verdugo, 2012).

Positive Psychology Interventions (PPI) are primarily intended to promote well-being in different spheres of life (Sin & Lyubomirsky, 2009). PPIs in sustainability context are intentional activities intended to cultivate positive cognitions, emotions, and actions in order to inculcate and maintain sustainability-centric behaviour. They do not aim to treat or fix the already existing harms to sustainability/climate.

For implementing PPI for sustainability, the foremost requirement is to develop a positive frame of mind and mindset to understand and appreciate the issues pertaining to sustainability, their dynamics and need for behavioural transformation. These Positive Sustainability Cognitions (PSC) can be established through Mindfulness-based Meditation focussed on nature sounds (chirping of birds, sound of water, and blowing of wind) which will establish human nature connect and promote gratefulness towards nature and its blessings to humankind. Further, cognitions can be strengthened through sustainability-centric capacity building with narratives and anecdotes highlighting best sustainability practices. Music and Art are some of the most powerful mediums to promote Positive Sustainability Emotions (PSE). There are collections available with sounds of wind, waves, water, and many other naturalistic sounds like sound of leaves, birds, bees, and other insects in the forest which can establish a positive emotional connect with natural environment. Similarly, visual media can also promote sustainability behaviour through documentaries and exhibiting best practices across the world, intended to develop positive cognitions and emotions. Sustainability Arts and Imagery is another field through which artists play a significant role in depicting the themes of sustainability and conservation of natural environment in a positive manner, having a large-scale impact, especially on young generation.

Sustainability Art
Art is a form of the expressions which converse acumens and open room for reflection in minds of human beings. It encompasses an expansive and diverse backdrop of unified creative practices including performing arts, literary arts, and the visual arts. There is an explicit connect of art and sustainability at global level to deploy the promotion of human rights, cultural diversity, well-being, and 'resolving the social and cultural challenges facing today's world'[11] Arts in varied forms have the potential to shape the sustainability behaviour of individuals in almost all socio-cultural settings. It acts as a protagonist to take a society towards ecological sustainability (Curtis et al., 2014). Since the eras of Plato and Aristotle, it has been acknowledged that people's beliefs, attitudes, and value systems in a civil society can be influenced by arts (Belfiore & Bennett, 2006) and these are the similar basis to influence pro-environmental behaviour (Kollmuss & Agyeman, 2002).

Sustainability art is now a well-recognized field and this connect between art and sustainability has been well documented by art sociologists (Kagan & Kirchberg, 2008). On the basis of interviews with 96 professionals in the arts and in the natural resource management sectors, Curtis et al. (2014) developed a model to pronounce the role of arts in shaping ecologically sustainable behaviour. The proposed pathways were engaging information communicating, empathy creation towards the natural environment and ecologically sustainable development. Artistic

[11] http://www.unesco.org/new/fileadmin/MULTIMEDIA/HQ/CLT/CLT/pdf/Arts_Edu_RoadMap_en.pdf.

images with the theme of sustainability can articulate a revelation for an ecologically sustainable society and the best practices besides provoking to look at creative ways of solving ecological problems and presenting sustainable solutions. Moldavanova (2013) proclaims that among environmental, economic, social, and cultural dimensions of sustainability, contribution of culture to promote sustainability in societies is key to an all-inclusive understanding of sustainability which can be beautifully reflected by embedded art forms and aesthetics.

To sum up, sustainability arts, mindful music, nature healing, landscape imaging, and nature narratives can together act as PPI for sustainability to generate positive cognitions and emotions leading to possible Positive Sustainability Actions (PSA), i.e. active involvement of people in pro-climate action and practising sustainable lifestyle.

Co-influencing the Eastern and the Western Thoughts

Asian/eastern and the western deliverance intersect in today's unified realm, with a convergence of diverse socio-cultural systems, offering its distinct forms of heightened consciousness, liberation, compassion, and assurances of larger good. The western deliverance primarily focuses on the transformation of social organizations, relationships, power structures, and the human minds budding within such liberating societies in the form of 'growing up' or transformative growth while the Asian/eastern liberation lays emphasis on the transformation of inner consciousness to awaken psycho-spiritual realities in the form of 'waking up' or contemplative growth. Both of them are intricately embedded in their own socio-culture milieu with their own distinctive strategy to blend with conservation and sustainability perspectives, but, at the same time, continue to cross-fertilize each other without entangling in their distinct metamorphoses and attributes (Snow, 2015), in order to derive universal benefits.

The scientifically proven western ecological knowledge is based on the understanding and appreciation for biodiversity on the planet, whereas the Asian/eastern indigenous knowledge relies on the understanding of how all living beings must be conserved and cared compassionately being the offsprings of common divine mother nature. Both the psycho-spiritual as well as rational-scientific understanding and appreciation for biodiversity are based on two diverse motivational paradigms and worldviews regarding the natural world, leading to contemplative as well as transformative development (Snow, 2015). However, they still can beautifully converge cognitive understanding of biodiversity of the west with compassionate and divine human–nature interface of the east and in a unified manner progressively heading towards pro-climate action and sustainability behaviour to assure climate and social justice. The field of inquiry blending the cognitive-spiritual framework of sustainability is contemplative neuroscience.

By linking the earning potential and living potential through spiritual intelligence and mindfulness, the west can learn compassion, contentment, and self-less service to the society while the eastern world, including the third world can learn the spirit of hard sustained work for development. Complimenting and supplementing the precious values across regions will develop the world for health and happiness.

Conclusion

Human response to changing climate and coping/adaptation efforts at micro- and macro-level will thoroughly depend on how we are able to positively manage the psycho-spiritual and social aspects for the benefit of future societies. Human psyche marked with spiritual intelligence, eco-consciousness, and mindfulness has the potential to manage anthropogenic causes of climate change in a way that is needed for a sustainable planet earth. Contemplative practices help in developing human resilience to climatic variability in a positive way. Exploring the spiritual dimensions of behaviour can be strongly instrumental in achieving CC Adaptation by enhancing the human coping potential. Mindfulness through spiritual intelligence can manifest in improving levels of climate change resilience (Salmabadi et al., 2016).

The greatest need of the hour is to enhance eco-consciousness as well as mindfulness by intentionally transforming the existing lifestyle and behaviour across the planet better suited to today's climate change and sustainability challenges. The adaptive strategies proposed by the government or institutions may have the limited success until and unless they are willingly and mindfully accepted by the communities. Hence, localized community-centric adaptive mechanisms are likely to be more effective. Co-management is a form of collective action to work together with a government agency to address some aspects which can be potentially threatening to the changing climate (Tompkins & Adger, 2004).

Development with environmental consciousness and mindfulness should be the target for which the world should be working together for preserving the precious human race and promoting global health and happiness. What is really needed is that people should have the cognitive understanding of the climate scenario from multidisciplinary perspectives. People should hear their inner voice to relook at their behavioural choices and convert their desirable choices into pro-climate/sustainability, which the earth is looking forward for a sustainable future. John Holmberg suggests that it will be possible only if people intend to move from *cruise mode*, i.e. management of routine sustainability goals, through steering and controlling unsustainable and uncalled-for actions, to *exploration mode*, i.e. exploring and leading proactive sustainability solutions through critical thinking, reflecting, and experimenting.[12] In-depth empirical research linking various disciplines of Human and Natural Sciences with cross-cutting methodologies is the need of the hour.

References

Abson, D. J., Von Wehrden, H., Baumgärtner, S., Fischer, J., Hanspach, J., Härdtle, W., Heinrichs, H., Klein, A., Lang, D., & Martens, P. (2014). Ecosystem services as a boundary object for sustainability. *Ecological Economics, 103*, 29–37.

Abson, D. J., Fischer, J., Leventon, J., Newig, J., Schomerus, T., Vilsmaier, U., ... Lang, D. J. (2017). Leverage points for sustainability transformation. *Ambio, 46*(1), 30–39.

Ajawani, J. (2017). Spiritual intelligence: A core ability behind psychosocial resilience. In U. Kumar (Ed.), *The Routledge international handbook of psychosocial resilience* (pp. 173–186). Routledge/Taylor & Francis Group.

Allison, E. (2016). Cultivating resilience for addressing ecological change. In *Embrace of the earth*.

Amel, E. L., Manning, C. M., & Scott, B. A. (2009). Mindfulness and sustainable behavior: Pondering attention and awareness as means for increasing green behavior. *Ecopsychology, 1*(1), 14–25. https://doi.org/10.1089/eco.2008.0005

Amram, Y. (2007). *The seven dimensions of spiritual intelligence: An ecumenical, grounded theory* (pp. 17–20).

---

[12] https://sttpaevents.com/session-summaries/.

Barbaro, N., & Pickett, S. M. (2016). Mindfully green: Examining the effect of connectedness to nature on the relationship between mindfulness and engagement in pro-environmental behavior. *Personality and Individual Differences, 93*, 137–142.
Belfiore, E., & Bennett, O. (2006). *Rethinking the social impact of the arts: A critical-historical review.* University of Warwick.
Broman, G. I., & Robèrt, K.-H. (2017). A framework for strategic sustainable development. *Journal of Cleaner Production, 140*, 17–31.
Brown, K. W., & Ryan, R. M. (2003). The benefits of being present: Mindfulness and its role in psychological well-being. *Journal of Personality and Social Psychology, 84*(4), 822–848. https://doi.org/10.1037/0022-3514.84.4.822
Carroll, J. (2016). Formless meditation and sustainability. In S. Dhiman & J. Marques (Eds.), *Spirituality and sustainability: New horizons and exemplary approaches.* Switzerland: Springer.
Chancellor, J., & Lyubomirsky, S. (2011). Happiness and thrift: When (spending) less is (hedonically) more. *Journal of Consumer Psychology, 21*(2), 131–138.
Curtis, D. J., Reid, N., & Reeve, I. (2014). Towards ecological sustainability: Observations on the role of the arts. *S.A.P.I.EN.S. Surveys and Perspectives Integrating Environment and Society, 7.1*, Article 7.1. http://journals.opened ition.org/sapiens/1655
Di Fabio, A. (2017). The psychology of sustainability and sustainable development for well-being in organizations. *Frontiers in Psychology, 8*(SEP), 1–7. https://doi.org/10.3389/fpsyg.2017.01534
Doty, J. R. (2016). *Into the magic shop: A neurosurgeon's true story of the life-changing magic of compassion and mindfulness.* Yellow Kite.
Dreyfus, G. (2011). Is mindfulness present-centred and non-judgmental? A discussion of the cognitive dimensions of mindfulness. *Contemporary Buddhism, 12*(1), 41–54.
Duerr, M. (2011). Assessing the state of contemplative practices in the US. In *Contemplative nation: How ancient practices are changing the way we live.*
Emmons, R. A. (2000). Is spirituality an intelligence? Motivation, cognition, and the psychology of ultimate concern. *International Journal of Phytoremediation, 10*(1), 3–26. https://doi.org/10.1207/S15327582IJPR1001_2
Ericson, T., Kjønstad, B. G., & Barstad, A. (2014). Mindfulness and sustainability. *Ecological Economics, 104*, 73–79. https://doi.org/10.1016/j.eco lecon.2014.04.007
Gärtner, C. (2013). Enhancing readiness for change by enhancing mindfulness. *Journal of Change Management, 13*(1), 52–68. https://doi.org/10.1080/ 14697017.2013.768433

Goralnik, L., & Marcus, S. (2020). Resilient learners, learning resilience: Contemplative practice in the sustainability classroom. *New Directions for Teaching and Learning, 2020*(161), 83–99.

Grossenbacher, P. G., & Quaglia, J. T. (2017). Contemplative cognition: A more integrative framework for advancing mindfulness and meditation research. *Mindfulness, 8*(6), 1580–1593. https://doi.org/10.1007/s12671-017-0730-1.Contemplative

Harding, S. (2006). *Animate earth: Science, intuition, and Gaia.* Chelsea Green Publishing.

Hedlund-de Witt, A. (2011). The rising culture and worldview of contemporary spirituality: A sociological study of potentials and pitfalls for sustainable development. *Ecological Economics, 70*(6), 1057–1065.

Hume, D. (1975). *A treatise of human nature* (2nd ed., L. A. Selby-Bigge, Eds.). Longmans Green & Co.

Ives, C. D., Freeth, R., & Fischer, J. (2020). Inside-out sustainability: The neglect of inner worlds. *Ambio, 49*(1), 208–217. https://doi.org/10.1007/s13280-019-01187-w

Kagan, S., & Kirchberg, V. (2008). *Sustainability: A new frontier for the arts and cultures: 2010.*

Khosravi, M., & Nikmanesh, Z. (2014). Relationship of spiritual intelligence with resilience and perceived stress. *Iranian Journal of Psychiatry, 8*(4), 52–56.

Koger, S. M. (2015). A burgeoning ecopsychological recovery movement. *Ecopsychology, 7*(4), 245–250.

Kollmuss, A., & Agyeman, J. (2002). Mind the gap: Why do people act environmentally and what are the barriers to pro-environmental behavior? *Environmental Education Research, 8*(3), 239–260.

Krok, D. (2008). The role of spirituality in coping: Examining the relationships between spiritual dimensions and coping styles. *Journal of Mental Health, Religion & Culture, 11*(7), 643–653.

Lama, D. (2001). *Ethics for the new millennium.* Penguin.

Levinson, D. B., Stoll, E. L., Kindy, S. D., Merry, H. L., & Davidson, R. J. (2014). A mind you can count on: Validating breath counting as a behavioral measure of mindfulness. *Frontiers in Psychology, 5,* 1202.

Liu, M., & Valente, E. (2018). *Psychology International| October 2018.* Psychology International. https://www.apa.org/international/pi/2018/10/mindfulness-climate-change

Mabsout, R. (2015). Mindful capability. *Ecological Economics, 112,* 86–97.

Manfredo, M. J., Bruskotter, J. T., Teel, T. L., Fulton, D., Schwartz, S. H., Arlinghaus, R., Oishi, S., Uskul, A. K., Redford, K., & Kitayama, S. (2017). Why social values cannot be changed for the sake of conservation. *Conservation Biology, 31*(4), 772–780.

Martin, J.-L., Maris, V., & Simberloff, D. S. (2016). The need to respect nature and its limits challenges society and conservation science. *Proceedings of the National Academy of Sciences, 113*(22), 6105–6112.

Martin-Breen, P., & Anderies, J. (2011). *Draft Background Paper*.

Moldavanova, A. (2013). Sustainability, ethics, and aesthetics. *The International Journal of Sustainability Policy and Practice, 8*(1), 109–120. https://doi.org/10.18848/2325-1166/CGP/v08i01/55356

Nguyen, N. C., & Bosch, O. J. (2013). A systems thinking approach to identify leverage points for sustainability: A case study in the Cat Ba Biosphere Reserve Vietnam. *Systems Research and Behavioral Science, 30*(2), 104–115.

Papenfuss, J., Merritt, E., Manuel-Navarrete, D., Cloutier, S., & Eckard, B. (2019). Interacting pedagogies: A review and framework for sustainability education. *Journal of Sustainability Education, 20,* 19.

Papenfuss, J. T. (2019). *Inside-out pedagogies: Transformative innovations for environmental and sustainability education*.

Rishi, P. (2009). Climate change and human psyche. *Environmental Issues: Behavioural Insights, 40*.

Robért, K.-H., Broman, G., & Basile, G., (2013). Analyzing the concept of planetary boundaries from a strategic sustainability perspective: How does humanity avoid tipping the planet? *Ecological Society, 18*(2), 5. https://doi.org/10.5751/ES-05336-180205

Salmabadi, M., Khamesan, A., Usefynezhad, A., & Sheikhipoor, M. (2016). The mediating role of spiritual intelligent in relationship of mindfulness and resilience. *Health Spiritual and Medical Ethics, 3*(3), 18–24. https://doi.org/10.4314/ajcem.v12i3

Schwartz, B., & Sharpe, K. E. (2006). Practical wisdom: Aristotle meets positive psychology. *Journal of Happiness Studies, 7*(3), 377–395.

Sin, N. L., & Lyubomirsky, S. (2009). Enhancing well-being and alleviating depressive symptoms with positive psychology interventions: A practice-friendly meta-analysis. *Journal of Clinical Psychology, 65*(5), 467–487.

Singh, H., Husain, T., Agnihotri, P., & Pandey, P. (2014). An ethnobotanical study of medicinal plants used in sacred groves of Kumaon Himalaya, Uttarakhand, India. *Elsevier, 154*(1), 98–108.

Sivilli, T. I., & Pace, T. W. (2014). The human dimensions of resilience: A theory of contemplative practices. *Garrison Institute*.

Snow, B. A. (2015). *Waking up and growing up: Two forms of human development*. Retrieved October 14, 2016.

Sol, J., & Wals, A. E. (2015). Strengthening ecological mindfulness through hybrid learning in vital coalitions. *Cultural Studies of Science Education, 10*(1), 203–214.

Srivastava, P. S. (2016). Spiritual intelligence: An overview. *International Journal of Multidisciplinary Research and Development, 3*(3), 224–227.

Tompkins, E. L., & Adger, W. N. (2004). Does adaptive management of natural resources enhance resilience to climate change? *Ecology and Society 9*, 10 (online). http://www.ecologyandsociety.org/vol9/iss2/art10

Vaughan, F. (2002). What is spiritual intelligence? *Journal of Humanistic Psychology, 42*(2), 16–33. https://doi.org/10.1177/0022167802422003

Verdugo, V. C. (2012). The positive psychology of sustainability. *Environment, Development and Sustainability, 14*(5), 651–666.

Wamsler, C. (2017). Stakeholder involvement in strategic adaptation planning: Transdisciplinarity and co-production at stake? *Environmental Science & Policy, 75*, 148–157.

Wamsler, C. (2019). Contemplative sustainable futures: The role of individual inner dimensions and transformation in sustainability research and education. In *Sustainability and the humanities* (pp. 359–373). Springer.

Wamsler, C., & Brink, E. (2018). Mindsets for sustainability: Exploring the link between mindfulness and sustainable climate adaptation. *Ecological Economics, 151*(April), 55–61. https://doi.org/10.1016/j.ecolecon.2018.04.029

Wamsler, C., Brossmann, J., Hendersson, H., Kristjansdottir, R., McDonald, C., & Scarampi, P. (2018). Mindfulness in sustainability science, practice, and teaching. *Sustainability Science, 13*(1), 143–162. https://doi.org/10.1007/s11625-017-0428-2

WWF. (2016). *Living planet report 2016: Risk and resilience in a new era*. Switzerland: Gland.

Zeng, X., Chiu, C., Wang, R., Oei, T., & Leung, F. Y. K. (2015). The effect of loving-kindness meditation on positive emotions: A meta-analytic review. *Frontiers in Psychology, 6*, 1693.

Zohar, D. (2012). *Spiritual intelligence: The ultimate intelligence*.

Zohar, D., & Marshall, I. (2001). *SQ-Spiritual Intelligence: The ultimate intelligence*.

CHAPTER 8

Conclusion: Looking Through a Behavior-Centric Prism

Human behaviours and lifestyles have left long-term cumulative influences on sustainability of ecological systems. Besides, there are enough tell-tale signs that are indicating how climate change is impacting our planet earth already. The way people have contributed to climate change through their unsustainable behavioural practices, climate change has primarily become a cause of concern. In general, there has been an overwhelming attention and significance attached to technological solutions for climate/sustainability promotion while less attention has been given to pro-climate behavioural transformation. However, it is behaviour-centric solutions at the level of individual, family, and community as recommended by psychological and behavioural sciences, which can be the starting points for any behavioural transformation to take place especially pertaining to climate change and sustainability.

Humanizing planet earth with care and compassion is what we are looking forward to, which is possible through an integration of natural, social, and behavioural sciences for managing climate change and sustainability. Sustainable behaviour not only entails responsible consumption practices such as carbon reduction for a better today but also means having environmental concerns for future generations as well. Effective solutions to climate change and sustainability can be made possible through a thorough understanding of human behavioural responses to climate change and sustainability. For humans to effectively cope and adapt to the changing climate scenario, they need to alter their behaviours and lifestyles as behavioural transformation plays a key role to ensure global well-being. In other words, behaviour transformation for sustainable consumption practices can help people go a long way in their fight to combat the battle of climate change.

Environmental concern and consciousness can trigger a forefront transformation towards sustainable behaviour irrespective of external rewards as a part of their intrinsic behaviour. Sustainable consumption will help in coping with prevalent social and financial challenges. COVID-19 pandemic has provided us with an opportunity to check the impact of human activities on the environment while introspecting our choices and behaviours. It has also changed the human–climate interface by affecting mobility, industrial, and construction activities and the consumption patterns along with providing us with an opportunity to and come with appropriate solutions through promoting positive behavioural change. These changes, if practised on a large scale, will push and motivate a large number of people to follow the best practices of the sustainable behavioural norms.

For ensuring the well-being of planet as well as its habitants, we have to move from '*wanting change*' to actually '*working for change*'—from the jungles of concrete to the lush green jungles; from the competitive race for unsustainable consumables to the peaceful collaboration in sharing resources; and the web of materialism to the alter of spiritualism. To extend compassionate care to inhabitants of planet earth is an integral part of sustainable behaviour. Although sustainability is not always cost-centric for people as well as corporate organizations, there are numerous long-term returns (tangible as well as intangible nature) attached to it. Effective adoption of Sustainable Development Goals (SDGs) by all individuals, societies, and nations can help in mitigation of climate change largely. It is an indisputable fact that human footprints are all over the planet and that climate change is largely anthropogenic in nature. It is the immense human interference with nature and natural resources along with human activities and lifestyle patterns which has resulted in climate change and there is enough research and scientific evidence to confirm this fact. It is very important to know what humans think of, feel about, and wish to do about this problem. However, mere cognitive awareness and risk perception about climate change is not sufficient and people are expected to be geared up for pro-climate action than their current levels of engagements. Through this book, we have demonstrated how different models of risk appraisal can be globally adapted to facilitate behavioural adaptation with climate change in diverse local to global perspectives.

Along with the policies for planet preservation and climate change, there is a need to look for alternative approaches and strategies in order to compensate for the apathetic human actions that have already caused irreparable damage to planet earth. Development, in this context means adequate and efficient utilization of locally available resources to create livelihood options as well as developing energy-efficient infrastructure, which is not detrimental to global climate. Additionally, it also requires people with a highly dedicated service-oriented mindset, who wish to take initiatives in the development of sustainable communities and society. This can be closely related to the true concept of sustainability in actual terms.

COVID-19 pandemic has given an opportunity to think of sustainability from a different perspective i.e. stay-at-home or work-from-home sustainability. This opportunity should be taken as a blessing in disguise as it gave us a chance to witness how keeping a check on human activities can have healing impacts for the environment while at the same time it pushed many of us into introspecting our behavioural choices. Learning to compromise with many of our behavioural choices and actions in order to fit in the 'new normal' made us realize that sustainability-linked choices may give us a desirable edge over common perception-based behaviour. Large-scale adoption of such behavioural choices can, in turn, inspire, push, and motivate a good number of people to inculcate best behavioural practices for managing climate change and sustainability.

Pandemic has not only changed the way we perceive our world but it has also given a massive chance to the corporate organizations to make their contributions to sustainability by way of their CSR activities and initiatives. In view of the new and upcoming challenges of the corporate world, society, and the environment at large, there has already been an imperative and pressing pre-requisite for the business world to willingly connect themselves with SDGs following the principles of corporate responsibility. The pandemic made us witness how corporates could strike a balance between profit and sustainability by strategic placement and alignment of their business operations along with CSR activities. Meaningful and insightful steps taken by various companies could help them draw strategic advantage out of the pandemic adversity along with making genuine, proactive as well as responsive contributions to society in the wake of pandemic and even beyond.

With the pandemic and associated loss to individual incomes as well as global economies, the disparities and inequalities in information access are further widened in developing countries. Already, in the context of developing countries, there is a special need to reverse the growing trends of migration towards urban settlements through more and more genuine developmental efforts on the part of government and private sector for raising the earning and living potential of rural communities and establishment of sustainable infrastructure. It cannot be ensured by just flowing national and international money or the CSR initiatives of corporate organizations. Since the scenario across the globe is differential in terms of resource availability, development potential as well as technological capacity/expertise, a more collaborative approach with equal responsibility sharing from both the developed and developing nations, besides encompassing the needs of diverse population groups is the need of the hour.

The 'digital divide' between the developing and developed countries has not only expanded the gap further between the 'front-runners' and 'back-seaters' of climate change and sustainability communication but it has already started to strain economies causing bends in socio-cultural and ecological environment globally. It is a clear-cut warning for assuming responsibility on the part of individuals, institutions as well as governance systems across the globe to redesign their relationships with each other and the natural environment.

Climate crisis across the globe cannot just be limited to explore climate solutions at the levels of corporate bodies and governance systems. Institutions and governance system alone cannot do anything without the wholehearted support and collaboration of human beings. As we all share a common responsibility for planet earth, a collective action is needed towards achievement of the targets of SDGs on behalf of all the countries. A positive, pro-climate, and sustainability-oriented mindset and action (even if it means sacrificing some of the luxuries of life) is desirable at the global level for which mass efforts of behavioural and social scientists around the globe are needed. A psychological and social process of understanding among diverse population groups with a vision for sustainability and a solution-oriented mindset can go a long way in this regard. Multidisciplinary approaches integrating natural and social sciences are warranted in this regard, to give policymakers, an insight into the various behavioural challenges and barriers in order to fortify policies for climate change and sustainability.

A great amount of hope can be attached to the concept of 'Frugality' which has a central and distinctive feature of a sustainable lifestyle necessitating reduced consumption and thereby reduced impact of behavioural practices on the 'availability and renewability of natural resources' (De Young, 1996). The concept is of great importance as it has the prospective to redefine our life choices and priorities by moving towards a more inclusive and sustainable society along with establishing harmony between living systems of planet earth. However, there is a need for transdisciplinary perspective to appreciate and apply this concept for the betterment of lives of a sizable population across the globe. The synergy between countries, disciplines, and professions can help embrace frugal innovation in a more meaningful manner as an enabling *'driver of progress in achieving sustainable solutions'*.

Another focal point is the linkage of climate change and sustainability to psycho-spiritual reality as it has the potential to strengthen behavioural adaptation to changing climatic scenario. Sustainability, when viewed from philosophical and psycho-spiritual basis, may be promoted through contemplative practices like mindfulness and spiritual intelligence. Hence, to face any climatic distress/adversity, a culmination of human, social, and technical vision is needed which is to be guided by spiritual wisdom. Such human-sustainability connect has the possibility to result in behavioural transformation for sustainability and climate change adaptation. For this, people need to know ways of enhancing the knowledge base in psycho-spiritual systems; ways to overcome the psychological barriers to climate change adaptation and sustainability management and developing the capacity to address climate change linked adversities through mindfulness and other socio-behavioural techniques. Furthermore, people and planet may also benefit from the application of psycho-spiritual concepts and an understanding of Positive Psychology concepts along with their implications thereby contributing positively to the environment and sustainable living.

Knowing that human life is inter-woven in a similar way as the global climate is, people have started thinking beyond their regional and national concerns to global concerns and coming together on the common platform surpassing all the national, cultural, social, religious, and ethnic boundaries. It is their responsibility to act in a way which ensures the sustainable future. This whole scenario revolves around the Indian thought *'VasudhevKutumbakam'* (in Sanskrit language) which means, 'the whole earth is like an extended family'. Everyone will have to collaborate and take care of each other when needed, share the available resources, celebrate the pleasures, and share the pains. This is the humanized approach for preserving the global climate for well-being and sustainable future.

People have to equitably share the global responsibility for climate change and sustainability through creation of multi-stakeholder milieu to proactively move towards SDG 13 (Climate Action) and take the sustainability agenda ahead. Systematic research regarding climate risk appraisal and perception for future extreme weather events in diverse geo-climatic regions/countries may help policymakers to integrate human dimensions in their documentations. Besides, micro-level adaptation efforts with the principal of co-management can also enhance the cumulative impact of adaptation efficacy. Active employee engagement as well as stakeholders' participation is crucial for impactful and effective CSR projects in environmental sustainability and pro-climate action domain.

The time has come when people will have to come out of their shell of indifference or inertia towards global environmental issues, transform their ecologically unsustainable behavioural practices, and raise their voices for pro-climate action. Besides climate talk, people will also move forward on climate walk for conserving the global climate as 'actions speak louder than words'. Tiny steps at local level put up together can cause global positive impact on climate. Everyone is capable enough for taking at least one such action/initiative at the local level, which can have a cumulative global impact. Micro-level steps, put in together, can convert into a macro-level strategy, ready for action. So, in spite of the climate change linked anxiety, generated by various reports, emotions can bring people in motion, provided they are channelized in the right direction. Hence, contemplative practices like mindfulness and spiritual intelligence can be instrumental to develop resilience for what cannot be changed and at the same time, motivate to adopt pro-climate concrete actions. The whole world is now geared up for that and the deliberations of mega climate summits prove it. What is needed is the people-centric effort at micro-level conjoining into the community-based actions under the umbrella of positive belief system and spirituality which prepares people for coping/adapting with any adverse scenario efficiently and at the same time acts as a buffer by providing a good support network in case of any climatic adversity.

REFERENCE

De Young, R. (1996). Some psychological aspects of reduced consumption behavior: The role of intrinsic satisfaction and competence motivation. *Environment and Behavior, 28*(3), 358–409.

Glossary of Keywords

Adaptation The act of modifying something or altering one's behaviour to make it apt for a new situation or action

Anthropogenic Caused by humans or their actions/activities. In the environmental context, it means resulting from the effect of human beings on environment

Biodiversity The diversity among living creatures including marine, terrestrial and other aquatic ecosystems together with the ecological complexes of which they are a part. This includes variability within species, between species and of ecosystems

Consumerism As a theory, it states that an increasing consumption of goods is economically desirable. It is an idea that growing consumption of goods and services is always a favourable outcome and that an individual's wellbeing and happiness lies profoundly on procurement of consumer goods and material assets

Consumption It is the act of consuming, as by use, decay, or destruction

Coping A set of actions taken up to deal with demanding, stressful or threatening situations. This is a kind of problem-solving *behaviour* which is aimed at bringing relief, reward, stillness, and stability

Dilapidation The condition of being old and in poor state or in the process of falling into decay/ being in poor shape

Eco/climate-Consciousness Refers to concern for the environment/climate. Individuals with this form of mindset feel directly connected to and involved with the natural world—plants, trees, animals, and insects as well as impacts of changing climate

Empathy "Putting oneself into the shoes of others". Ability to gauge other people's emotions along with the ability to imagine what someone else might be thinking or feeling; *Empathy* is the first step towards sympathy

Experiential Concerning or resulting from experience and learning can be based on observation and trial-and-error

Empirical *Empirical* is something that is based solely on experiment or experience

Frugality It is the quality or state of being frugal in which careful management of material resources and especially money is being practised

Lockdown A constraint or confinement policy for public to stay where they are and this is generally due to particular risks to themselves or to others because of free movement or interaction with others

Materialism A predisposition to think of material possessions and physical comfort as more significant than spiritual tenets or values

Mindfulness A basic human ability to be fully present and of being aware or conscious of where we are and what we're doing while at the same time not overly reactive or carried away by what's going on around us

Mitigation Lowering the risk of impairment from the occurrence of any undesirable event i.e. to minimize the degree of any harm or loss

Pandemic It is more widespread than an epidemic and typically occurs over a wide geographical area like several countries or continents and particularly affects a substantial proportion of the population

Perception It refers to see things as "whole" i.e. the way in which something is seen, understood, or interpreted

Psychodynamics It is the connection of the unconscious and conscious mental and emotional forces that results in personality and motivation of an individual. Early childhood and their impacts on behaviour and mental states are particularly important

Resilience It is a person's ability to bounce back after an intense setback and to recover from or adjust easily to adversity or change

Risk Appraisal An assessment or evaluation of the possibilities that a future event or happening may occur

Sustainability Meeting the requirements and needs of the present without compromising the capacity of future generations to meet their needs

Sustainability communication Sustainability communication is essential for promoting and managing stakeholder expectations in terms of social requests; it is also a potent tool for assessing how sustainable firms' commitment to socio-environmental issues is developed

Sustainable management The coming together of business and *sustainability* through the practice of *managing* a business's impact on people, planet, and profit in such a way that all three can prosper in the future

Transcendence The act of rising above one's present state to a superior state

Urban Trance It refers to the state that we are in when we get excessively engrossed in what we are doing and shut the world around us

Urbanization It is the concentration of inhabitants in bigger urban settlements of a said territory and in terms of the rising compactness of population inside urban settlements

Watershed management The watershed is a zone of land which drains water into a particular receiving waterbody like lake, river etc. Watershed management is the process of applying both land use practices and water management practices to safeguard and improve water quality along with other natural resources within the scope of a watershed in an all-inclusive manner

Index

A
Aesthetics, 192
Affective, 7, 20, 28, 29, 32, 35, 41, 48, 63, 68–70, 72, 124, 147, 148, 151, 175
Anthropogenic, 1, 3, 26, 64, 126, 193, 200
Asceticism, 86
Asian, 80, 85, 86, 192
Attention, 27, 28, 49, 58, 62, 63, 67, 71, 85, 109, 138, 141, 142, 147, 148, 152, 176, 185, 199
Austerity, 86

B
Behaviour, 1–3, 7–14, 18, 20, 25–27, 29, 31, 33–35, 38, 39, 41, 48, 61–63, 66, 68, 73, 79, 80, 85, 93, 99, 100, 106, 113, 114, 121, 132, 137, 138, 140, 143–152, 154–156, 159–162, 175, 178, 180, 181, 183, 185, 186, 188–191, 193, 199–201

Behavioural change, 2, 7–10, 18, 20, 29, 49, 69, 138, 139, 141–143, 146, 149, 154–157, 159, 161, 162, 200
Behavioural choice, 83, 91–93, 139, 140, 159, 194, 201
Behavioural compulsion, 83
Behavioural restraint, 176, 188, 189
Behavioural science, 13, 26, 36, 61, 137, 140, 142, 144, 161, 162, 170, 189, 199
Behavioural transformation, 28, 38, 42, 62, 69, 132, 138, 139, 141, 143, 146, 147, 149, 154, 155, 159, 161, 171, 175, 177, 181, 188, 190, 199, 200, 203
Behaviour impact cycles, 27
Belief systems, 1, 7, 139, 141, 150, 151, 171, 173, 177, 180, 204
Buddhism, 86, 87

© The Editor(s) (if applicable) and The Author(s), under exclusive license to Springer Nature Singapore Pte Ltd. 2022
P. Rishi, *Managing Climate Change and Sustainability through Behavioural Transformation*, Sustainable Development Goals Series, https://doi.org/10.1007/978-981-16-8519-4

C

Climate change (CC), 1–5, 10–12, 14, 15, 17, 18, 20, 21, 25–31, 33–39, 41, 42, 47–50, 59, 60, 62, 64–74, 82, 94, 106, 110, 113, 124, 126, 137–149, 154–157, 159, 161, 162, 169–173, 175–177, 180–185, 188–190, 193, 199–204

Climate Change Communication (C3), 64–69, 72, 73

Climate fatigue, 64

Cognitive, 7, 20, 26–28, 31–33, 35, 36, 41, 49, 62–64, 66, 69, 70, 124, 140, 149, 151, 154, 159, 174, 178, 181, 184–186, 193, 194, 200

Community-based adaptive management, 142

Conscious, 5, 30, 36, 49, 63, 73, 148, 149, 154, 157, 158

Conservation, 5, 7, 15, 17, 19, 33, 79, 86, 88, 90, 100, 106, 113, 117, 125, 151, 156, 174, 175, 177, 191, 192

Consumptive practices, 7, 13, 14, 63, 80, 81, 83, 184

Contemplative neuroscience, 193

Contemplative practices, 162, 170, 171, 175–177, 186, 188, 189, 193, 203, 204

Corporate social responsibility (CSR), 51, 53, 74, 84, 95, 101, 105–132, 201, 202, 204

COVID-19, 10, 13, 14, 18, 20, 47, 48, 59, 69–73, 98, 126–130, 200, 201

D

Dayalbagh, 11, 87, 90, 91, 93, 157
Digital divide, 47, 48, 202

E

Eco-city, 81

Ecologically sustainable, 27, 175, 191, 192

Ecological resilience, 176, 186

Ecological sustainability, 81, 111, 113, 191

Eco-village, 90

Emissions, 3, 4, 8, 11, 12, 14–17, 19, 20, 57, 59, 84, 125, 152, 155–157, 160

Empathy, 70, 144, 176, 179, 186

Employee engagement, 55, 121, 122, 124, 131, 204

Energy, 4, 11, 39, 59, 68, 72, 73, 80, 90, 98–100, 115, 125, 131, 156–160, 180, 189, 201

Environment, 1, 4, 5, 7, 9–11, 14–20, 25, 27, 33, 40–42, 48–50, 53–55, 57–59, 61–64, 69–71, 74, 81, 83–85, 88–95, 99–101, 105, 106, 108–117, 121, 123–126, 131, 137, 140, 142, 144, 146, 147, 149–152, 154, 157, 159, 161, 171, 173, 175–177, 180–183, 186–192, 194, 199–204

Environmental attitudes, 142, 145

Equity, 80, 100, 113–116, 183, 188

Ethos, 83

F

Frugality, 80–96, 99–101, 143, 203

G

Globalization, 3, 79, 106

Governance, 15, 25, 36, 48, 88, 106, 115, 117, 137, 140, 169, 202

Greenwashing, 58, 73

H

Happiness, 81, 82, 86, 88, 89, 91, 93, 94, 100, 175, 187, 189, 193, 194
Human–climate interface, 1, 14, 25, 200

I

Inclusive, 2, 49, 69, 72, 73, 79, 80, 100, 101, 105, 112, 117, 125, 126, 203
India, 8, 11, 17, 18, 30, 39, 41, 48, 53, 54, 56, 81, 83–85, 90, 91, 93, 95, 97, 98, 101, 107, 112, 113, 116–118, 126, 129, 130, 148, 157, 160, 177
Innovation, 5, 33, 62, 82, 83, 95–99, 101, 108, 115, 116, 139, 141, 159, 203

J

Jugaad, 11, 83, 95, 97–99

K

Knowledge, 26–28, 30, 35, 41, 69, 120, 124, 126, 131, 138, 146, 150, 151, 153, 156, 160, 170–172, 174, 180, 181, 192, 203

L

Lifestyles, 10, 30, 80–82, 86, 87, 91, 92, 94, 137, 147, 154, 156, 158, 177, 183, 189, 193, 200, 203
Locus of control (LOC), 144

M

Meditation, 176, 177, 186, 190
Mental processes, 28, 29, 31, 105, 151
Millennials, 57, 67, 82
Minimalistic lifestyle, 81

N

Neuroplasticity, 187
Neuroscience, 187
Neurotransmitters, 179
Non-judgmental, 182

P

Pandemic, 10, 13–20, 48, 59, 70–73, 128–132, 200–202
Perceived risk, 33, 154, 159
Perception, 2, 5, 21, 25, 26, 28, 29, 32–36, 38, 41, 56, 61, 72, 106, 113, 122, 123, 143, 151, 154, 155, 185, 201, 204
Pluralistic, 49, 69
Poverty alleviation, 115
Protection Motivation Theory (PMT), 27, 33, 38
Psyche, 28, 50, 67, 189, 193
Psychological distance, 39, 71, 72, 159
Psycho-spiritual, 170, 171, 173, 175, 192, 193, 203

R

Radhasoami sect, 90, 91
Redemption, 177
Reflective thinking, 49
Reflexive modernity, 93
Resilience, 28, 39–41, 86, 142, 176, 181, 185, 187, 188, 190, 193, 204
Responsible consumption, 7, 39, 79, 115, 151, 199

Risk appraisal, 21, 26, 27, 31, 32, 36, 39–41, 201, 204
Risk perception, 26, 28–31, 33–39, 172, 200

S
Self-acceptance, 177
Self-efficacy, 38, 41, 100, 154
Shared value, 50, 52, 53
Social contract, 126
Social discourse, 49, 50
Social Network Analysis (SNA), 123, 124
Social norms, 9, 20, 35, 39, 154
Social Process Reengineering (SPR), 118–120, 123, 124
Social Representation Theory (SRT), 28, 30, 31
Social systems, 1, 2, 26, 115, 116, 118, 186
Soft skills, 120, 121, 124
Spirituality, 81, 175, 177, 178, 180, 181, 204
Stakeholders, 49, 51–57, 60, 61, 73, 105, 108, 109, 111, 117, 118, 121, 123–125, 130, 131, 149, 204
Stereotypes, 49
Subjective well-being, 89
Supra-materialism, 93, 94
Sustainability, 1, 3, 5, 7–11, 14, 17–20, 39, 48–63, 69–71, 73, 74, 79–82, 84, 85, 91, 93–96, 98–101, 108, 109, 111, 112, 114–118, 121, 124–126, 130–132, 137–144, 146–151, 153, 155, 157–159, 162, 169–171, 173–194, 199–204
Sustainability Art, 191
Sustainability communication, 49–52, 60, 62, 69
Sustainability crisis, 171, 175, 183, 184, 190
Sustainability management, 1–3, 15, 116, 170–172, 203
Sustainability performance, 51, 55, 58, 59
Sustainability strategy, 51
Sustainable consumption behaviour (SCB), 143–145, 150, 151, 162
Sustainable Development Goals (SDGs), 5, 12, 16, 20, 47, 51, 53–55, 72–74, 79, 80, 99, 101, 106, 108, 111–113, 115–117, 125–127, 131, 140, 157, 158, 161, 200–202
Sustainable materialism, 93
Sustainable value chain, 56

T
Tranquillity, 81, 176
Transactional, 106, 107, 123
Transcendence, 83, 176–178
Transformational, 106, 108, 123, 140
Transformational adaptation, 139, 140
Transformative development, 193

U
Urbanization, 3, 20, 137

V
Value, 5, 7, 8, 10, 20, 25, 33, 35, 42, 49, 50, 52–54, 57, 59, 62, 63, 66, 69, 72–74, 85, 88, 90, 91, 94, 100, 101, 106, 113, 116, 117, 122, 128, 138–141, 145, 150, 151, 153, 156, 157, 159, 170, 173, 175, 177, 183, 186, 188, 191, 193
Value shift, 175
Vocational skills, 116
Voluntary simplicity, 81, 82, 86, 93

Printed in the United States
by Baker & Taylor Publisher Services